The Meme Machine

Susan Blackmore is a Reader in Psychology at the University of the West of England, Bristol, where she lectures on the psychology of consciousness. Dr Blackmore's research interests include near-death experiences, the effects of meditation, why people believe in the paranormal, evolutionary psychology, and the theory of memetics. She is the current Perrott-Warrick Researcher, studying psychic phenomena in borderline states of consciousness, and has received the Distinguished Skeptic's Award from CSICOP, the Committee for the Scientific Investigation of Claims of the Paranormal. Susan Blackmore writes for several magazines, has an occasional column in the *Independent* newspaper, and is a frequent contributor and presenter on radio and television.

The Meme Machine

SUSAN BLACKMORE

OXFORD
UNIVERSITY PRESS

Great Clarendon Street, Oxford OX2 6DP

Oxford University Press is a department of the University of Oxford
and furthers the University's aim of excellence in research, scholarship,
and education by publishing worldwide in

Oxford New York

Athens Auckland Bangkok Bogotá Bombay Buenos Aires Calcutta
Cape Town Chennai Dar es Salaam Delhi Florence Hong Kong Istanbul
Karachi Kuala Lumpur Madras Melbourne Mexico City Mumbai
Nairobi Paris São Paulo Singapore Taipei Tokyo Toronto Warsaw

and associated companies in Berlin Ibadan

Oxford is a registered trade mark of Oxford University Press

Published in the United States
by Oxford University Press Inc., New York

First published 1999
First issued as an Oxford University Press paperback 2000

British Library Cataloguing in Publication Data

(Data available)

Library of Congress Cataloging in Publication Data

(Data available)

ISBN 0-19-286212-X

Typeset by Joshua Associates Ltd., Oxford
Printed in Great Britain
by Cox & Wyman Ltd.
Reading, Berks.

For Adam

Foreword

by Richard Dawkins

As an undergraduate I was chatting to a friend in the Balliol College lunch queue. He regarded me with increasingly quizzical amusement, then asked: 'Have you just been with Peter Brunet?' I had indeed, though I couldn't guess how he knew. Peter Brunet was our much loved tutor, and I had come hotfoot from a tutorial hour with him. 'I thought so', my friend laughed. 'You are talking just like him; your voice sounds exactly like his.' I had, if only briefly, 'inherited' intonations and manners of speech from an admired, and now greatly missed, teacher. Years later, when I became a tutor myself, I taught a young woman who affected an unusual habit. When asked a question which required deep thought, she would screw her eyes tight shut, jerk her head down to her chest and then freeze for up to half a minute before looking up, opening her eyes, and answering the question with fluency and intelligence. I was amused by this, and did an imitation of it to divert my colleagues after dinner. Among them was a distinguished Oxford philosopher. As soon as he saw my imitation, he immediately said: 'That's Wittgenstein! Is her surname ____ by any chance?' Taken aback, I said that it was. 'I thought so', said my colleague. 'Both her parents are professional philosophers and devoted followers of Wittgenstein.' The gesture had passed from the great philosopher, via one or both of her parents to my pupil. I suppose that, although my further imitation was done in jest, I must count myself a fourth-generation transmitter of the gesture. And who knows where Wittgenstein got it?

The fact that we unconsciously imitate others, especially our parents, those in quasi-parental roles, or those we admire, is familiar enough. But is it really credible that imitation could become the basis of a major theory of the evolution of the human mind and the explosive inflation of the human brain, even of what it means to be a conscious self? Could imitation have been the key to what set our ancestors apart from all other animals? I would never have thought so, but Susan Blackmore in this book makes a tantalisingly strong case.

Imitation is how a child learns its particular language rather than some other language. It is why people speak more like their own parents than like other people's parents. It is why regional accents, and on a longer timescale separate languages, exist. It is why religions persist along family lines rather than being chosen afresh in every generation. There is at least a superficial analogy to the longitudinal transmission of genes down generations, and to the horizontal transmission of genes in viruses. Without prejudging the issue of whether the analogy is a fruitful one, if we want even to talk about it we had better have a name for the entity that might play the role of gene in the transmission of words, ideas, faiths, mannerisms and fashions. Since 1976, when the word was coined, increasing numbers of people have adopted the name 'meme' for the postulated gene analogue.

The compilers of the *Oxford English Dictionary* operate a sensible criterion for deciding whether a new word shall be canonised by inclusion. The aspirant word must be commonly used without needing to be defined and without its coinage being attributed whenever it is used. To ask the metamemetic question, how widespread is 'meme'? A far from ideal, but nevertheless easy and convenient method of sampling the meme pool, is provided by the World Wide Web and the ease with which it may be searched. I did a quick search of the Web on the day of writing this, which happened to be 29 August 1998. 'Meme' is mentioned about half a million times, but that is a ridiculously high figure, obviously confounded by various acronyms and the French *même.* The adjectival form 'memetic', however, is genuinely exclusive, and it clocked up 5042 mentions. To put this number into perspective, I compared a few other recently coined words or fashionable expressions. Spin doctor (or spin-doctor) gets 1412 mentions, dumbing down 3905, docudrama (or docu-drama) 2848, sociobiology 6679, catastrophe theory 1472, edge of chaos 2673, wannabee 2650, zippergate 1752, studmuffin 776, post-structural (or poststructural) 577, extended phenotype 515, exaptation 307. Of the 5042 mentions of memetic, more than 90 per cent make no mention of the origin of the word, which suggests that it does indeed meet the *OED*'s criterion. And, as Susan Blackmore tells us, the *Oxford English Dictionary* now does contain the following definition:

> **meme** An element of a culture that may be considered to be passed on by non-genetic means, esp. imitation.

Further searching of the Internet reveals a newsgroup talking shop, 'alt.memetics', which has received about 12 000 postings during the past year. There are on-line articles on, among many other things, 'The New Meme', 'Meme, Counter-meme', 'Memetics: a Systems Metabiology',

'Memes, and Grinning Idiot Press', 'Memes, Metamemes and Politics', 'Cryonics, religions and memes', 'Selfish Memes and the evolution of cooperation', and 'Running down the Meme'. There are separate Web pages on 'Memetics', 'Memes', 'The C Memetic Nexus', 'Meme theorists on the Web', 'Meme of the week', 'Meme Central', 'Arkuat's Meme Workshop', 'Some pointers and a short introduction to memetics', 'Memetics Index' and 'Meme Gardening Page'. There is even a new religion (tongue-in-cheek, I *think*), called the 'Church of Virus', complete with its own list of Sins and Virtues, and its own patron saint (Saint Charles Darwin, canonised as 'perhaps the most influential memetic engineer of the modern era') and I was alarmed to discover a passing reference to 'St Dawkin'.

Susan Blackmore's book is preceded by two others entirely devoted to the subject of memes and both good in their different ways: Richard Brodie's *Virus of the Mind: The New Science of the Meme*, and Aaron Lynch's *Thought Contagion: How Belief Spreads through Society*. Most significant of all, the distinguished philosopher Daniel Dennett has adopted the idea of the meme, building it in as a cornerstone of his theory of mind, as developed in his two great books *Consciousness Explained*, and *Darwin's Dangerous Idea*.

Memes travel longitudinally down generations, but they travel horizontally too, like viruses in an epidemic. Indeed, it is largely horizontal epidemiology that we are studying when we measure the spread of words like 'memetic', 'docudrama' or 'studmuffin' over the Internet. Crazes among schoolchildren provide particularly tidy examples. When I was about nine, my father taught me to fold a square of paper to make an origami Chinese junk. It was a remarkable feat of artificial embryology, passing through a distinctive series of intermediate stages: catamaran with two hulls, cupboard with doors, picture in a frame, and finally the junk itself, fully seaworthy or at least bathworthy, complete with deep hold, and two flat decks each surmounted by a large, square-rigged sail. The point of the story is that I went back to school and infected my friends with the skill, and it then spread around the school with the speed of the measles and pretty much the same epidemiological time-course. I do not know whether the epidemic subsequently jumped to other schools (a boarding school is a somewhat isolated backwater of the meme pool). But I do know that my father himself originally picked up the Chinese Junk meme during an almost identical epidemic at the same school 25 years earlier. The earlier virus was launched by the school matron. Long after the old matron's departure, I had reintroduced her meme to a new cohort of small boys.

Before leaving the Chinese Junk, let me use it to make one more point. A favourite objection to the meme/gene analogy is that memes, if they exist at all, are transmitted with too low fidelity to perform a gene-like role in any realistically Darwinian selection process. The difference between high-fidelity genes and low-fidelity memes is assumed to follow from the fact that genes, but not memes, are digital. I am sure that the details of Wittgenstein's mannerism were far from faithfully reproduced when I imitated my pupil's imitation of her parents' imitation of Wittgenstein. The form and timing of the tic undoubtedly mutated over the generations, as in the childhood game of Chinese Whispers (Americans call it Telephone).

Suppose we assemble a line of children. A picture of, say, a Chinese junk is shown to the first child, who is asked to draw it. The drawing, but not the original picture, is then shown to the second child, who is asked to make her own drawing of it. The second child's drawing is shown to the third child, who draws it again, and so the series proceeds until the twentieth child, whose drawing is revealed to everyone and compared with the first. Without even doing the experiment, we know what the result will be. The twentieth drawing will be so unlike the first as to be unrecognisable. Presumably, if we lay the drawings out in order, we shall note some resemblance between each one and its immediate predecessor and successor, but the mutation rate will be so high as to destroy all semblance after a few generations. A trend will be visible as we walk from one end of the series of drawings to the other, and the direction of the trend will be degeneration. Evolutionary geneticists have long understood that natural selection cannot work unless the mutation rate is low. Indeed, the initial problem of overcoming the fidelity barrier has been described as the Catch-22 of the Origin of Life. Darwinism depends on high-fidelity gene replication. How then can the meme, with its apparently dismal lack of fidelity, serve as quasi-gene in any quasi-Darwinian process?

It is not always as dismal as you think and, as Susan Blackmore insists, high fidelity is not necessarily synonymous with digital. Suppose we set up our Chinese Whispers Chinese Junk game again, but this time with a crucial difference. Instead of asking the first child to copy a drawing of a junk, we teach her, by demonstration, to make an origami model of a junk. When she has mastered the skill and made her own junk, the first child is asked to turn round to the second child and teach him how to make one. So the skill passes down the line to the twentieth child. What will be the result of this experiment? What will the twentieth child produce, and what shall we observe if we lay the twenty efforts out in order along the ground? I have not done it, but I will make the following confident prediction, assuming that we run the experiment many times on different groups of twenty children. In several of the experiments, a child somewhere along the line will forget some crucial step in the skill taught him by the previous child, and the line of phenotypes will suffer an abrupt macromutation which will presumably then be copied to the end of the line, or until another discrete mistake is made. The end result of such mutated lines will not bear any resemblance to a Chinese junk at all. But in a good number of experiments the skill will correctly pass all along the line, and the twentieth junk will be no worse and no better, on average, than the first junk. If we then lay the twenty junks out in order, some will be more perfect than others, but imperfections will not be copied on down the line. If the fifth child is ham-fisted and makes a clumsily asymmetrical or floppy junk, his quantitative errors will be corrected if the sixth child happens to be more dextrous. The twenty junks will not exhibit a progressive deterioration in the way that the twenty drawings of our first experiment undoubtedly would.

Why? What is the crucial difference between the two kinds of experiment? It is this: inheritance in the drawing experiment is Lamarckian (Blackmore calls it 'copying-the-product'). In the origami experiment it is Weismannian (Blackmore's 'copying-the-instructions'). In the drawing experiment, the phenotype in every generation is also the genotype – it is what is passed on to the next generation. In the origami experiment, what passes to the next generation is not the paper phenotype but a set of instructions for making it. Imperfections in the execution of the instructions result in imperfect junks (phenotypes) but they are not passed on to future generations: they are non-memetic. Here are the first five instructions in the Weismannian meme line of instructions for making a Chinese junk:

1. Take a square sheet of paper and fold all four corners exactly into the middle.
2. Take the reduced square so formed, and fold one side into the middle.
3. Fold the opposite side into the middle, symmetrically.
4. In the same way, take the rectangle so formed, and fold its two ends into the middle.
5. Take the small square so formed, and fold it backwards, exactly along the straight line where your last two folds met.

. . . and so on, through 20 or 30 instructions of this kind. These instructions, though I would not wish to call them digital, are potentially of very high fidelity, just as if they were digital. This is because they all make reference to idealised tasks like 'fold the four corners exactly into the middle'. If the paper is not exactly square, or if a child folds ineptly so that, say, the first corner overshoots the middle and the fourth corner undershoots it, the junk that results will be inelegant. But the next child in the line will not copy the error, for she will assume that her instructor *intended* to fold all four corners into the exact centre of a perfect square. The instructions are self-normalising. The code is error-correcting. Plato would enjoy it: what passes down the line is an ideal essence of junk, of which each actual junk is an imperfect approximation.

The instructions are more effectively passed on if verbally reinforced, but they can be transmitted by demonstration alone. A Japanese child could teach an English one, though neither has a word of the other's language. In the same way, a Japanese master carpenter could convey his skills to an equally monoglot English apprentice. The apprentice would not copy obvious mistakes. If the master hit his thumb with a hammer, the apprentice would correctly guess, even without understanding the Japanese expletive '** **** ** !', that he meant to hit the nail. He would not make a Lamarckian copy of the precise details of every hammer blow, but copy instead the inferred Weismannian instruction: drive the nail in with as many blows of your hammer as it takes your arm to achieve the same idealised end result as the master has achieved with his – a nail head flush with the wood.

I believe that these considerations greatly reduce, and probably remove altogether, the objection that memes are copied with insufficient high fidelity to be compared with genes. For me, the quasi-genetic inheritance of language, and of religious and traditional customs, teaches the same lesson. Another objection, discussed, like the first, in Susan Blackmore's illuminating chapter on 'Three problems with memes' is that we do not know what memes are made of or where they reside. Memes have not yet found their Watson and Crick; they even lack their Mendel. Where genes are to be found in precise locations on chromosomes, memes presumably exist in brains, and we have even less chance of seeing one than of seeing a gene (though, in an article referred to by Blackmore, the neurobiologist Juan Delius had pictured his conjecture of what a meme might look like). As with genes, we track memes through populations by their phenotypes. The 'phenotype' of the Chinese junk meme is made of paper. With the exception of 'extended phenotypes', such as beaver dams and caddis larva houses, the phenotypes of genes are normally parts of living bodies. Meme phenotypes seldom are.

But it can happen. To return to my school again, a Martian geneticist, visiting the school during the morning cold bath ritual, would have unhesitatingly diagnosed an 'obvious' genetic polymorphism. About 50 per cent of the boys were circumcised and 50 per cent were not. The boys, incidentally, were highly conscious of the polymorphism and we classified ourselves into Roundheads versus Cavaliers (I have recently read of another school in which the boys even organised themselves into two football teams along the same lines). It is, of course, not a genetic but a memetic polymorphism. But the Martian's mistake is completely understandable; the morphological discontinuity is of exactly the kind that one normally expects to find produced by genes.

In England at that time, infant circumcision was a medical whim, and the Roundhead/Cavalier polymorphism at my school probably owed less to longitudinal transmission than to differing fashions in the various hospitals where we happened to have been born – horizontal memetic transmission, yet again. But through most of history circumcision has been longitudinally transmitted as a badge of religion (of *parents*' religion I hasten to point out, for the unfortunate child is normally too young to *know* his own religious mind). Where circumcision is religiously or traditionally based (the barbaric custom of female circumcision always is), the transmission will follow a longitudinal pattern of heredity, very similar to the pattern for true genetic transmission, and often persisting for many generations. Our Martian geneticist would have to work quite hard to discover that no genes are involved in the genesis of the roundhead phenotype.

The Martian geneticist's eyes would also pop out on stalks (assuming they were not on stalks to begin with) at the contemplation of certain styles of clothing and hairdressing, and their inheritance patterns. The black skull-capped phenotype shows a marked tendency towards longitudinal transmission from father to son (or it may be from maternal grandfather to grandson), and there is clear linkage to the rarer pigtail-plaited sideburn phenotype. Behavioural phenotypes such as genuflecting in front of crosses, and facing east to kneel five times per day, are inherited longitudinally too, and are in strongly negative linkage disequilibrium with each other and with the previously mentioned phenotypes, as is the red-dot-on-forehead phenotype, and the saffron robes/shaven head linkage group.

Genes are accurately copied and transmitted from body to body, but some are transmitted at greater frequency than others – by definition they are more successful. This is natural selection, and it is the explanation for most of what is interesting and remarkable about life. But is there a similar meme-based natural selection? Perhaps we can use the Internet again to investigate natural selection among memes? As it happens, around the time the word 'meme' was coined (actually a little later), a rival synonym, 'culturgen', was proposed. Today, culturgen is mentioned twenty times on the World Wide Web, compared with memetic's 5042. Moreover, of those twenty, seventeen also mention the source of the word, falling foul of the *Oxford English Dictionary*'s criterion. Perhaps it is not too fanciful to imagine a Darwinian struggle between the two memes (or culturgens), and it is not totally silly to ask why one of them was so much more successful. Perhaps it is because meme is a monosyllable similar to gene, which therefore lends itself to quasi-genetic sub-coinings: meme pool (352), memotype (58), memeticist (163), memeoid (or memoid) (28), retromeme (14), population memetics (41), meme complex (494), memetic engineering (302) and metameme (71) are all listed in the 'Memetic Lexicon' at *http://www.lucifer.com/virus/memlex.html§MEME* (the numbers in parentheses count the mentions of each word on the Web on my sampling day). Culturgen-based equivalents would be more obvious but less snappy. Or the success of meme against culturgen may have been initially just a non-Darwinian matter of chance – memetic drift (85) – followed by a self-reinforcing positive feedback effect ('unto every one that hath shall be given, and he shall have abundance; but from him that hath not shall be taken away even that which he hath', Matthew 25: 29).

I have mentioned two favourite objections to the meme idea: memes have insufficient copying fidelity, and nobody really knows what a meme physically is. A third is the vexed question of how large a unit deserves the name 'meme'. Is the whole Roman Catholic Church one meme, or should we use the word for one constituent unit such as the idea of incense or transubstantiation? Or for something in between? Susan Blackmore gives due attention to such questions, but she rightly concentrates on a more constructive approach, developing the positive explanatory power of the 'memeplex' – an abbreviation which she prefers over the full 'coadapted meme complex', and I shall be surprised if in time her book does not bring about a Darwinian reversal of their numerical fortunes (today, 20 and 494, respectively).

Memes, like genes, are selected against the background of other memes in the meme pool. The result is that gangs of mutually compatible memes – coadapted meme complexes or memeplexes – are found cohabiting in individual brains. This is not because selection has chosen them as a group, but because each separate member of the group tends to be favoured when its environment happens to be dominated by the others. An exactly similar point can be made about genetic selection.

Every gene in a gene pool constitutes part of the environmental background against which the other genes are naturally selected, so it's no wonder natural selection favours genes that 'cooperate' in building those highly integrated and unified machines called organisms. Biologists are sharply divided into those for whom this logic is as clear as daylight, and those (even some very distinguished ones) who just do not understand it – who naïvely trot out the obvious cooperativeness of genes and unitariness of organisms as though they somehow counted against the 'selfish gene' view of evolution. Susan Blackmore not only understands it, she explains the matter with unusual clarity and goes on to apply the lesson with equal clarity and force to memes. By analogy with coadapted gene complexes, memes, selected against the background of each other, 'cooperate' in mutually supportive memeplexes – supportive within the memeplex but hostile to rival memeplexes. Religions may be the most convincing examples of memeplexes but they are by no means the only ones. Susan Blackmore's treatment is, as ever, provocative and revealing.

I believe a sufficient case has been made that the analogy between memes and genes is persuasive and that the obvious objections to it can be satisfactorily answered. But can the analogy do useful work? Can it lead us to powerful new theories that actually explain anything important? This is where Susan Blackmore really comes into her own. She warms us up with some fascinating vignettes which get us used to the memetic style of reasoning. Why do we talk so much? Why can't we stop thinking? Why do silly tunes buzz round our heads and torment us into insomnia? In every case she begins her response in the same way: 'Imagine a world full of brains, and far more memes than can possibly find homes. Which memes are more likely to find a safe home and get passed on again?' The answer comes back readily enough, and our understanding of ourselves is enriched. She pushes on, with patience and skill applying the same method to deeper and more exacting problems: What is language for? What attracts us to our mates? Why are we so good to each other? Did memes drive the rapid, massive, and peculiar evolutionary expansion of the human brain? Along the way, she shows how the theory of memes can throw light on particular areas where she has special expertise from her academic career as a psychologist and sceptical investigator of the paranormal: superstition and near-death experience.

In the end, showing greater courage and intellectual *chutzpah* than I have ever aspired to, she deploys her memetic forces in a brave – do not think foolhardy until you have read it – assault on the deepest questions of all: What is a self? What am I? Where am I? (famous questions posed by Daniel Dennett long before he became the philosophical mentor of all meme theorists). What of consciousness, creativity and foresight?

I am occasionally accused of having backtracked on memes; of having lost heart, pulled in my horns, had second thoughts. The truth is that my first thoughts were more modest than some memeticists, including perhaps Dr Blackmore, might have wished. For me, the original mission of the meme was negative. The word was introduced at the end of a book which otherwise must have seemed entirely devoted to extolling the selfish gene as the be-all and end-all of evolution, the fundamental unit of selection, the entity in the hierarchy of life which all adaptations could be said to benefit. There was a risk that my readers would misunderstand the message as being *necessarily* about genes in the sense of DNA molecules. On the contrary, DNA was incidental. The real unit of natural selection was any kind of *replicator*, any unit of which copies are made, with occasional errors, and with some influence or power over their own probability of replication. The genetic natural selection identified by neo-Darwinism as the driving force of evolution on this planet was only a special case of a more general process that I came to dub 'Universal Darwinism'. Perhaps we would have to go to other planets in order to discover any other examples. But perhaps we did not have to go that far. Could it be that a new kind of Darwinian replicator was even now staring us in the face? This was where the meme came in.

I would have been content, then, if the meme had done its work of simply persuading my readers that the gene was only a special case: that its role in the play of Universal Darwinism could be filled by any entity in the universe answering to the definition of Replicator. The original didactic purpose of the meme was the negative one of cutting the selfish gene down to size. I became a little alarmed at the number of my readers who took the meme more positively as a theory of human culture in its own right – either to criticise it (unfairly, given my original modest intention) or to carry it far beyond the limits of what I then thought justified. This was why I may have seemed to backtrack.

But I was always open to the possibility that the meme might one day be developed into a proper hypothesis of the human mind, and I did not know how ambitious such a thesis might turn out to be. Any theory deserves to be given its best shot, and that is what Susan Blackmore has given the theory of the meme. I do not know whether she will be judged too ambitious in this enterprise, and I would even fear for her if I did not know her redoubtable qualities as a fighter. Redoubtable she is, and hard-nosed too, but at the same time her style is light and personable. Her thesis undermines our most cherished illusions (as she would see them) of individual identity and personhood, yet she comes across as the kind of individual person you would wish to know. As one reader I am grateful for the courage, dedication and skill she has put into her difficult task of memetic engineering, and I am delighted to recommend her book.

Preface

This book owes its existence to an illness. In September 1995 I caught a nasty virus, and struggled to keep working until I was finally forced to give up and take to my bed. I stayed there for many months, unable to walk more than a few steps, unable to talk for more than a few minutes, unable to use my computer – in fact unable to do anything but read and think.

During this time I began on my pile of 'urgent books I must read this week' which had long been oppressing me. One of them was Dan Dennett's latest book *Darwin's Dangerous Idea*. At about the same time one of my PhD students, Nick Rose, wrote me an essay on 'Memes and Consciousness'. Somehow the meme meme got to me. I had read Dawkins's *The Selfish Gene* many years before but, I suppose, had dismissed the idea of memes as nothing more than a bit of fun. Suddenly I realised that here was a powerful idea, capable of transforming our understanding of the human mind – and I hadn't even noticed it. I then read everything I could find on memes. Since I had to refuse all invitations to give lectures, take part in television programmes, go to conferences, or write papers, I could devote myself properly to the study of memes.

Most of the ideas in this book came to me while I was lying in bed during those months, especially between January and March 1996. As I gradually got better I began to make extensive notes. Some two years after I first became ill I was well enough to work again, and decided to keep on saying no to all those invitations, and to write this book instead.

I would like to thank the illness for making it possible, and my children Emily and Jolyon for not, apparently, minding that their mother was uselessly lying in bed all the time. I would like to thank my partner Adam Hart-Davis for not only looking after me when I was ill, but for encouraging my enthusiasm for memes in every way possible and for putting 'the book' first.

Dan Dennett was one of the first to hear my ideas and I thank him for his 'avuncular advice'. Several people helped greatly by reading earlier drafts of all or part of the book. They are Richard Dawkins, Dan Dennett,

Derek Gatherer, Adam Hart-Davis, Euan MacPhail, Nick Rose, and my editor Michael Rodgers who has given me much sound advice and encouragement. Helena Cronin helped enormously by inviting me to lecture on memes and putting me in touch with many helpful critics. Finally I would like to thank the Perrott–Warrick Fund for their financial support for the research on sleep paralysis and the paranormal discussed in Chapter 14. Without all this help, these particular memes would never have come together.

Bristol SJB
October 1998

Contents

CHAPTER 1

Strange creatures

We humans are strange creatures. There is no doubt that our bodies evolved by natural selection just as other animals' did. Yet we differ from all other creatures in many ways. For a start we speak. We believe ourselves to be the most intelligent species on the planet. We are extraordinarily widespread and extremely versatile in our ways of making a living. We wage wars, believe in religions, bury our dead and get embarrassed about sex. We watch television, drive cars and eat ice cream. We have had such a devastating impact upon the ecosystems of our planet that we appear to be in danger of destroying everything on which our lives depend. One of the problems of being a human is that it is rather hard to look at humans with an unprejudiced eye.

On the one hand, we are obviously animals comparable with any others. We have lungs, hearts and brains made of living cells; we eat and breathe and reproduce. Darwin's theory of evolution by natural selection can successfully explain how we, along with the rest of life on this planet, came to be here, and why we all share so many characteristics. On the other hand, we behave quite differently from other animals. Now that biology has so successfully explained much of our similarity with other creatures we need to ask the opposite question. What makes us so different? Could it be our superior intelligence, our consciousness, our language, or what?

A common answer is that we are simply more intelligent than any other species. Yet the notion of intelligence is extremely slippery, with interminable arguments about how to define it, how to measure it, and to what extent it is inherited. Research in artificial intelligence (AI) has provided some nice surprises for those who thought they knew what makes human intelligence so special.

In the early days of AI, researchers thought that if they could teach a computer to play chess they would have reproduced one of the highest forms of human intelligence. In those days the idea that a computer could ever play well, let alone beat a Grand Master, was unthinkable. Yet now most home computers come with passable chess programmes already installed, and in 1997 the program *Deep Blue* beat World Champion Garry Kasparov, ending unquestioned human supremacy at the game.

Computers may not play chess in the same way as humans, but their success shows how wrong we can be about intelligence. Clearly, what we thought were human beings' most special capabilities may not be.

Quite the opposite goes for some apparently quite unintelligent things like cleaning the house, digging the garden or making a cup of tea. Time and again AI researchers have tried to build robots to carry out such tasks and been defeated. The first problem is that the tasks all require vision. There is a popular (though possibly apocryphal) story about Marvin Minsky at MIT (the Massachusetts Institute of Technology) that he once gave his graduate students the problem of vision as a summer project. Decades later the problem of computer vision is still just that – a problem. We humans can see so effortlessly that we cannot begin to imagine how complex the process has to be. And in any case, this kind of intelligence cannot distinguish us from other animals because they can see too.

If intelligence does not provide simple answers perhaps consciousness might. Many people believe that human consciousness is unique and is responsible for making us human. Yet scientists cannot even define the term 'consciousness'. Everyone knows what their own consciousness is like but they cannot share that knowledge with anyone else. This troublesome fact – the subjectivity of consciousness – may explain why for most of this century the whole topic of consciousness was more or less banned from scientific discussion. Now at last it has become fashionable again, but scientists and philosophers cannot even agree on what an explanation of consciousness would look like. Some say that the 'Hard Problem' of subjectivity is quite different from any other scientific problem and needs a totally new kind of solution, while others are sure that when we fully understand brain function and behaviour the problem of consciousness will have disappeared.

Some people believe in the existence of a human soul or spirit that transcends the physical brain and explains human uniqueness. With the decline in religious belief fewer and fewer people intellectually accept that view, yet most of us continue to think of ourselves as a little conscious 'me' inside our brain; a 'me' who sees the world, makes the decisions, directs the actions and has responsibility for them.

As we shall see later, this view has to be wrong. Whatever the brain is doing it does not seem to need help from an extra, magical self. Various parts of the brain carry on their tasks independently of each other and countless different things are always going on at once. We may feel as though there is a central place inside our heads in to which the sensations come and from which we consciously make the decisions. Yet this place simply does not exist. Clearly, something is very wrong with our ordinary view of our conscious selves. From this confused viewpoint we cannot say with certainty that other animals are not conscious, nor that consciousness is what makes us unique. So what does?

What makes us different?

The thesis of this book is that what makes us different is our ability to imitate.

Imitation comes naturally to us humans. Have you ever sat and blinked, or waved, or 'goo gooed', or even just smiled, at a baby? What happens? Very often they blink too, or wave, or smile back at you. We do it so easily, even as an infant. We copy each other all the time. Like seeing, it comes so effortlessly that we hardly think about it. We certainly do not think of it as being something very clever. Yet, as we shall see, it is fantastically clever.

Certainly, other animals do not take naturally to it. Blink, or wave, or smile at your dog or cat and what happens? She might purr, wag her tail, twitch, or walk away, but you can be pretty sure she will not imitate you. You can teach a cat, or rat, to beg neatly for its food by progressively rewarding it, but you cannot teach it by demonstrating the trick yourself – nor can another cat or rat. Years of detailed research on animal imitation has led to the conclusion that it is extremely rare (I shall return to this in Chapter 4). Though we may think of mother cats as teaching their kittens to hunt, or groom, or use the cat door, they do not do it by demonstration or imitation. Parent birds 'teach' their babies to fly more by pushing them out of the nest and giving them the chance to try it than by demonstrating the required skills for them to copy.

There is a special appeal to stories of animals copying human behaviour, and pet owners are fond of such tales. I read on the Internet about a cat who learned to flush the toilet and soon taught a second cat the same trick. Now the two of them sit together on the cistern flushing away. A more reliable anecdote was told by Diana Reiss, a psychologist at Rutgers University. She works with bottlenose dolphins, who are known to be able to copy vocal sounds and artificial whistles, as well as simple actions (Bauer and Johnson 1994; Reiss and McCowan 1993). She trained the dolphins by giving them fish as a reward and also by a 'time out' procedure for punishment. If they did the wrong thing she would walk away from the water's edge and wait for one minute before returning to the pool. One day she threw a fish to one of the dolphins but had accidentally left on some spiky bits of fin. Immediately the dolphin turned, swam away, and waited for a minute at the other side of the pool.

That story touched me because I could not help thinking of the dolphins as *understanding* the action, as having intelligence and consciousness and intentionality like ours. But we cannot even define these things, let alone be sure that the dolphin was using them in this apparent act of reciprocation. What we can see is that it *imitated* Dr Reiss in an appropriate way. We are so oblivious to the cleverness of imitation that we do not even notice how rare it is in other animals and how often we do it ourselves.

Perhaps more telling is that we do not have separate words for radically different kinds of learning. We use the same word 'learning' for simple association or 'classical conditioning' (which almost all animals can do), for learning by trial and error or 'operant conditioning' (which many animals can do), and for learning by imitation (which almost none can do). I want to argue that the supreme ease with which we are capable of imitation, has blinded us to this simple fact – that *imitation* is what makes us special.

Imitation and the meme

When you imitate someone else, something is passed on. This 'something' can then be passed on again, and again, and so take on a life of its own. We might call this thing an idea, an instruction, a behaviour, a piece of information . . . but if we are going to study it we shall need to give it a name.

Fortunately, there is a name. It is the 'meme'.

The term 'meme' first appeared in 1976, in Richard Dawkins's best-selling book *The Selfish Gene*. In that book Dawkins, an Oxford zoologist, popularised the increasingly influential view that evolution is best understood in terms of the competition between genes. Earlier in the twentieth century, biologists had blithely talked about evolution occurring for the 'good of the species' without worrying about the exact mechanisms involved, but in the 1960s serious problems with this view began to be recognised (Williams 1966). For example, if a group of organisms all act for the good of the group then one individual who does not can easily exploit the rest. He will then leave more descendants who in turn do not act for the group, and the group benefit will be lost. On the more modern 'gene's eye view', evolution may *appear* to proceed in the interests of the individual, or for the good of the species, but in fact it is all driven by the competition between genes. This new viewpoint provided a much more powerful understanding of evolution and has come to be known as 'selfish-gene theory'.

We must be absolutely clear about what 'selfish' means in this context. It does not mean genes *for* selfishness. Such genes would incline their carriers to act selfishly and that is something quite different. The term 'selfish' here means that the genes act only for themselves; their only interest is their own replication; all they want is to be passed on to the next generation. Of course, genes do not 'want' or have aims or intentions in the same way as people do; they are only chemical instructions that can be copied. So when I say they 'want', or are 'selfish' I am using a shorthand, but this shorthand is necessary to avoid lengthy explanations. It will not lead us astray if we remember that genes either *are* or *are not* successful at getting passed on into the next generation. So the shorthand 'genes want *x*' can always be spelled out as 'genes that do *x* are more likely to be passed on'. This is the only power they have – replicator power. And it is in this sense that they are selfish.

Dawkins also introduced the important distinction between 'replicators' and their 'vehicles'. A replicator is anything of which copies are made, including 'active replicators' whose nature affects the chances of their being copied again. A vehicle is the entity that interacts with the environment, which is why Hull (1988*a*) prefers the term 'interactors' for a similar idea. Vehicles or interactors carry the replicators around inside them and protect them. The original replicator was presumably a simple self-copying molecule in the primeval soup, but our most familiar replicator now is DNA. Its vehicles are organisms and groups of organisms that interact with each other as they live out their lives in the seas or the air, the forests or fields. Genes are the selfish replicators that drive the evolution of the biological world here on earth but Dawkins believes there is a more fundamental principle at work. He suggested that wherever it arises, anywhere in the universe, 'all life evolves by the differential survival of replicating entities' (1976, p. 192). This is the foundation for the idea of Universal Darwinism; the application of Darwinian thinking way beyond the confines of biological evolution.

At the very end of the book he asked an obvious, if provocative, question. Are there any other replicators on our planet? The answer, he claimed, is 'Yes'. Staring us in the face, although still drifting clumsily about in its primeval soup of culture, is another replicator – a unit of imitation.

> We need a name for the new replicator, a noun that conveys the idea of a unit of cultural transmission, or a unit of *imitation*. 'Mimeme' comes from a suitable Greek root, but I want a monosyllable that sounds a bit like 'gene'. I hope my classicist friends will forgive me if I abbreviate mimeme to *meme*.

As examples, he suggested 'tunes, ideas, catch-phrases, clothes fashions, ways of making pots or of building arches'. He mentioned scientific ideas that catch on and propagate themselves around the world by jumping from brain to brain. He wrote about religions as groups of memes with a high survival value, infecting whole societies with belief in a God or an afterlife. He talked about fashions in dress or diet, and about ceremonies, customs and technologies – all of which are spread by one person copying another. Memes are stored in human brains (or books or inventions) and passed on by imitation.

In a few pages, Dawkins laid the foundations for understanding the evolution of memes. He discussed their propagation by jumping from brain to brain, likened them to parasites infecting a host, treated them as physically realised living structures, and showed how mutually assisting memes will gang together in groups just as genes do. Most importantly, he treated the meme as a replicator in its own right. He complained that many of his colleagues seemed unable to accept the idea that memes would spread for their own benefit, independently of any benefit to the genes. 'In the last analysis they wish always to go back to "biological advantage"' to answer questions about human behaviour. Yes, he agreed, we got our brains for biological (genetic) reasons but now we have them a new replicator has been unleashed. 'Once this new evolution begins, it will in no necessary sense be subservient to the old' (Dawkins 1976, pp. 193–4). In other words, memetic evolution can now take off without regard to its effects on the genes.

If Dawkins is right then human life is permeated through and through with memes and their consequences. Everything you have learned by imitation from someone else is a meme. But we must be clear what is meant by the word 'imitation', because our whole understanding of memetics depends on it. Dawkins said that memes jump from 'brain to brain via a process which, in the broad sense, can be called imitation' (1976, p. 192). I will also use the term 'imitation' in the broad sense. So if, for example, a friend tells you a story and you remember the gist and pass it on to someone else then that counts as imitation. You have not precisely imitated your friend's every action and word, but something (the gist of the story) has been copied from her to you and then on to someone else. This is the 'broad sense' in which we must understand the term 'imitation'. If in doubt, remember that something must have been copied.

Everything that is passed from person to person in this way is a meme. This includes all the words in your vocabulary, the stories you know, the skills and habits you have picked up from others and the games you like to play. It includes the songs you sing and the rules you obey. So, for example, whenever you drive on the left (or the right!), eat curry with lager or pizza and coke, whistle the theme tune from *Neighbours* or even shake hands, you are dealing in memes. Each of these memes has evolved in its own unique way with its own history, but each of them is using your behaviour to get itself copied.

Take the song 'Happy Birthday to You'. Millions of people – probably thousands of millions of people the world over – know this tune. Indeed, I only have to write down those four words to have a pretty good idea that you may soon start humming it to yourself. Those words affect you, probably quite without any conscious intention on your part, by stirring up a memory you already possess. And where did that come from? Like millions of other people you have acquired it by imitation. Something, some kind of information, some kind of instruction, has become lodged in all those brains so that now we all do the same thing at birthday parties. That something is what we call the meme.

Memes spread themselves around indiscriminately without regard to whether they are useful, neutral, or positively harmful to us. A brilliant new scientific idea, or a technological invention, may spread because of its usefulness. A song like Jingle Bells may spread because it sounds OK, though it is not seriously useful and can definitely get on your nerves. But some memes are positively harmful – like chain letters and pyramid selling, new methods of fraud and false doctrines, ineffective slimming diets and dangerous medical 'cures'. Of course, the memes do not care; they are selfish like genes and will simply spread if they can.

Remember that the same shorthand applies to memes as to genes. We can say that memes are 'selfish', that they 'do not care', that they 'want' to propagate themselves, and so on, when all we mean is that successful memes are the ones that get copied and spread, while unsuccessful ones do not. This is the sense in which memes 'want' to get copied, 'want' you to pass them on and 'do not care' what that means to you or your genes.

This is the power behind the idea of memes. To start to think memetically we have to make a giant flip in our minds just as biologists had to do when taking on the idea of the selfish gene. Instead of thinking of our ideas as our own creations, and as working for us, we have to think of them as autonomous selfish memes, working only to get themselves copied. We humans, because of our powers of imitation, have become just the physical 'hosts' needed for the memes to get around. This is how the world looks from a 'meme's eye view'.

Meme fear

This is a scary idea indeed. And perhaps that is why the word 'meme' is so often written with inverted commas around it, as though to apologise for using it. I have even seen eminent lecturers raise both hands and tweak them above their ears when forced to say 'meme' out loud. Gradually, the word has become more generally known, and has even been added to the *Oxford English Dictionary*. There are discussion groups and a *Journal of Memetics* on the Internet, and the idea almost seems to have acquired a cult following in cyberspace. But in academia it has not yet been so successful. A perusal of some of the best recent books on human origins, the evolution of language and evolutionary psychology shows that the word does not appear at all in most of them ('meme' is not in the indexes of Barkow *et al.* 1992; Diamond 1997; Dunbar 1996; Mithen 1996; Pinker 1994; Mark Ridley 1996; Tudge 1995; Wills 1993; Wright 1994). The idea of memes seems extremely relevant to these disciplines, and I want to argue that it is time for us to take on board the notion of a second replicator at work in human life and evolution.

One of the problems with the idea of memes is that it strikes at our deepest assumptions about who we are and why we are here. This is always happening in science. Before Copernicus and Galileo, people believed they lived at the centre of the universe in a world created especially for them by God. Gradually, we had to accept not only that the sun does not revolve around the earth, but that we live on some minor little planet in an ordinary galaxy in a vast universe of other galaxies.

A hundred and forty years ago Darwin's theory of evolution by natural selection provided the first plausible mechanism for evolution without a designer. People's view of their own origin changed from the biblical story of special creation in the image of God, to an animal descended from an apelike ancestor – a vast leap indeed, and one that led to much ridicule and fanatical opposition to Darwin. Still – we have all coped with that leap and come to accept that we are animals created by evolution. However, if memetics is valid, we will have to make another vast leap in accepting a similar evolutionary mechanism for the origin of our minds and our selves.

....

What will determine whether the theory of memes is worth having or not? Although philosophers of science argue over what makes a scientific theory valid, there are at least two commonly agreed criteria, and I will use these in judging memetics. First, a theory must be able to explain things better than its rival theories; more economically or more comprehensively. And second, it must lead to testable predictions that turn out to be correct. Ideally, those predictions should be unexpected ones – things that no one would have looked for if they were not starting from a theory of memetics.

My aim in this book is to show that many aspects of human nature are explained far better by a theory of memetics than by any rival theory yet available. The theory starts only with one simple mechanism – the competition between memes to get into human brains and be passed on again. From this, it gives rise to explanations for such diverse phenomena as the evolution of the enormous human brain, the origins of language, our tendency to talk and think too much, human altruism, and the evolution of the Internet. Looked at through the new lens of the memes, human beings look quite different.

Is the new way better? It seems obviously so to me, but I expect that many people will disagree. This is where the predictions come in. I shall try to be as clear as I can in deriving predictions and showing how they follow from memetic theory. I may speculate and even, at times, leap wildly beyond the evidence, but as long as the speculations can be tested then they can be helpful. In the end, the success or failure of these predictions will decide whether memes are just a meaningless metaphor or the grand new unifying theory we need to understand human nature.

CHAPTER 2

Universal Darwinism

Darwin's theory of evolution by natural selection is, to my mind, the most beautiful in all of science. It is beautiful because it is so simple and yet its results are so complex. It is counter-intuitive and hard to grasp but once you have seen it the world is transformed before your eyes. There is no longer any need for a grand designer to explain all the complexity of the living world. There is just a stark and mindless procedure by which we have all come about – beautiful but scary.

I want to spend most of this chapter explaining the theory. The problem is that this beautifully simple idea is often misunderstood. Perhaps its very simplicity makes people think there must be something more to it, or that they have missed the point when they have actually grasped it. Evolution by natural selection is very, very simple and not at all obvious.

Darwin explained the basic principle in his great work *On the Origin of Species by Means of Natural Selection*, first published in 1859. Before that time many people had been impressed by the relationships between organisms, and by progressions in the fossil record, and had speculated about evolution. Among them were Charles's grandfather Erasmus Darwin, and Jean-Baptiste de Lamarck. However, no one had described a plausible mechanism by which evolution might work, and this was Darwin's great contribution.

He reasoned that if living creatures vary (as they certainly do) and if, due to their geometric increase in numbers, there is at certain times a struggle for life (which cannot be disputed), then it would be most extraordinary if there were not some variation that was useful to a creature's welfare. The individuals with these characteristics will then have the best chance of being 'preserved in the struggle for life' and will produce offspring with the same characteristics. This was the principle he called 'natural selection'.

Darwin's argument requires three main features: variation, selection and retention (or heredity). That is, first there must be variation so that not all creatures are identical. Second, there must be an environment in which not all the creatures can survive and some varieties do better than others. Third, there must be some process by which offspring inherit characteristics from their parents. If all these three are in place then any characteristics that are positively useful for survival in that environment must tend to increase. Put into Richard Dawkins's language, if there is a replicator that makes imperfect copies of itself only some of which survive, then evolution simply *must* occur. This *inevitability* of evolution is part of what makes Darwin's insight so clever. All you need is the right starting conditions and evolution just has to happen.

The evolutionary algorithm

The American philosopher Daniel Dennett (1995) has described the whole evolutionary process as an algorithm, that is, a mindless procedure which, when followed, must produce an outcome. Nowadays we are used to the idea of algorithms, although Darwin, Wallace and other early evolutionists would not have been. Many of the things we do are based on algorithms, whether it is adding up sums, dialling a telephone number or even making a cup of tea. Our interactions with machines are particularly algorithmic and the prevalence of machines makes it easier for us to think this way – take a cup, put it under the spout, choose the drink, put in the right amount of money, press the button, take the cup out – if you do the right steps in the right order then the result is a cup of cappuccino, do it wrong and you have a mess on the floor. The computer programs that hold our medical records or run the graphics in our computer games are all algorithms, as are the ways we interface with word processors and financial packages.

Algorithms are 'substrate-neutral', meaning they can run on a variety of different materials. A human with a pencil and paper, a hand-cranked adding machine, and a digital computer can all follow the same algorithm for some mathematical procedure and come to the same answer. The substrate does not matter – only the logic of the procedure does. In the case of Darwin's own argument the substrate was living creatures and a biological environment, but as Dennett points out his logic would apply equally to any system in which there was heredity, variation, and selection. This, again, is the idea of Universal Darwinism.

Algorithms are also completely mindless. If a system is set up so that it follows a given procedure then it does not also need a little mind, or extra-something, inside to make it work. It just must mindlessly happen. This is why Dennett describes Darwin's theory as 'a scheme for creating Design out of Chaos without the aid of Mind' (1995, p. 50). The design simply must come about when millions of creatures, over millions of years, produce more offspring than can survive. The ones that live do so because they are better adapted to the environment in which they find themselves. They then pass on their characteristics to their offspring and so it goes on. The environment itself is constantly changing because of all these developments, and so the process is never static.

Algorithms must always produce the same result if they start from the same point. This seems to suggest that, if evolution follows an algorithm, its results must be predetermined and predictable. This is not the case, and chaos theory explains why not. There are many simple processes, like dripping taps or moving gases, or the path drawn out by a swinging pendulum, which are chaotic. They follow simple and mindless algorithms but their end results are complex, chaotic and unpredictable. Beautiful shapes and patterns can emerge, but although the *kind* of pattern may be repeatable, the detail cannot be predicted without running the procedure right through. And since chaotic systems can be highly sensitive to initial starting conditions, a tiny difference at the beginning may lead to an entirely different outcome. Evolution is like this.

The complexity theorist Stuart Kauffman also likens the evolution of life to an incompressible computer algorithm. We cannot predict exactly how it will all unfold and can only 'stand back and watch the pageant'. We can, however, 'find deep and beautiful laws governing that unpredictable flow' (Kauffman 1995, p. 23).

We can now see that even if evolution is only following a simple algorithm, it is a chaotic system and its outcome can be incredibly complex. Moreover, the results cannot be predicted without running it – and it is only being run once. We can do experiments to test predictions of the theory, but we cannot rerun the evolution of life on earth to see whether it might go a different way next time. There is no next time. Until we find life on other planets there is only this once.

Many interesting arguments remain: such as just how much pattern and order inevitably springs up in the universe even without selection; the role of historical accidents in shaping the path of life, and whether evolution will always tend to produce certain kinds of thing, such as wormlike creatures with a mouth at the front, symmetrical animals with pairs of legs, or eyes or sex. Their resolution will help our understanding of evolution enormously but none of this really matters for grasping the basic principle of the evolutionary algorithm. When this algorithm gets going the inevitable result is that design is created out of nowhere – but we cannot predict exactly what sort of design it will be. Evolution emphatically did not have to end up with us. It had to end up with something more than it started with – and that something just happens to be this world with us in it.

Is there progress in evolution? Gould (1996*a*) famously argues there is not, but I think he has a concept of progress that I do not share. He is right to rule out progress *towards* anything. This is the whole point of Darwin's inspiration – and what makes his theory so beautiful – there is no master plan, no end point, and no designer. But of course there is progress in the sense that we now live in a complex world full of creatures of all kinds and a few billion years ago there was only a primeval soup. Although there is no generally accepted measure for this complexity, there is no doubt that the variety of organisms, the number of genes in individual organisms, and their structural and behavioral complexity have all increased (Maynard Smith and Szathmáry 1995). Evolution uses its own products to climb upon.

Dawkins (1996*a*) describes this as 'Climbing Mount Improbable' – as time goes on natural selection inches up the gentle slopes to reach the heights of ever more improbable creations, and when there are strong selection pressures, progress may be maintained for many generations. Dennett describes the progress as 'lifting in Design Space', the crane or wedge of natural selection very slowly, and by tiny steps, finds and accumulates good design tricks by building on the efforts of all the earlier climbing. In this sense, then, there is progress.

This progress is not necessarily steady or always increasing. There are long periods of stasis between periods of rapid change. Also, some animals, like crocodiles, stay the same for long periods, while others change rapidly. And sometimes millions of years of accumulated design are suddenly wiped out, as when the dinosaurs became extinct. Some people believe that we humans are in the process of obliterating as much biodiversity as was lost in that previous extinction. If we do, then the evolutionary algorithm will start its creative work again on whatever is left.

All this creativity depends on replicator power. The selfish replicators get copied, and they do this willy-nilly so long as they have the machinery and building blocks they need for that copying. They have no foresight, they do not look ahead or have plans or schemes in mind. They just get copied. In the process some do better than others – some obliterate others – and in this way evolutionary design comes about.

These, then, are some of the general principles that apply to any theory of evolution. If memes are really replicators and can sustain an evolutionary process then all these principles must apply and we should be able to build a theory of memetics on this basis. So are they? We can now ask two important questions – What are the criteria for being a replicator? Does the meme fulfil those criteria?

Memes as replicators

For something to count as a replicator it must sustain the evolutionary algorithm based on variation, selection and retention (or heredity). Memes certainly come with *variation* – stories are rarely told exactly the same way twice, no two buildings are absolutely identical, and every conversation is unique – and when memes are passed on, the copying is not always perfect. As the psychologist, Sir Frederic Bartlett (1932) showed in the 1930s, a story gets a bit embellished or details are forgotten every time it is passed on. There is memetic *selection* – some memes grab the attention, are faithfully remembered and passed on to other people, while others fail to get copied at all. Then, when memes are passed on there is *retention* of some of the ideas or behaviours in that meme – something of the original meme must be retained for us to call it imitation or copying or learning by example. The meme therefore fits perfectly into Dawkins's idea of a replicator and into Dennett's evolutionary algorithm.

Let us consider the example of a simple story. Have you heard the one about the poodle in the microwave? An American lady, so the story goes, used to wash her poodle and dry it in the oven. When she acquired a brand new microwave oven, she did the same thing, bringing the poor dog to a painful and untimely death. Then she sued the manufacturers for not providing a warning 'Do not dry your poodle in this oven' – and won!

This story has spread so widely that millions of people in Britain have heard it – but they might have heard another version, like the 'cat in the microwave' version, or the 'Chihuahua in the microwave'. Perhaps Americans have an equivalent version in which the woman is from New York or Kansas City. This is an example of an 'urban myth', a story that takes on a life of its own regardless of its truth, value or importance. This story is probably untrue but truth is not a necessary criterion for a successful meme. If a meme can spread, it will.

Stories like this are clearly inherited – millions of people cannot have suddenly made up the same story by chance, and the way the changes creep in can be used to demonstrate where a story originated and how it spread. There is clearly variation – not everyone has heard the same version even though the basic story is recognisable. Finally, there is selection – millions of people tell millions of stories every day but most are completely forgotten and only very few achieve urban-myth status.

Where do new memes come from? They come about through variation and combination of old ones – either inside one person's mind, or when memes are passed from person to person. So, for example, the poodle story is concocted out of language that people already know and ideas they already have, put together in new ways. They then remember it and pass it on, and variations occur in the process. And the same is true of inventions, songs, works of art, and scientific theories. The human mind is a rich source of variation. In our thinking we mix up ideas and turn them over to produce new combinations. In our dreams we mix them up even more, with bizarre – and occasionally creative – consequences. Human creativity is a process of variation and recombination.

In thinking about thinking we should remember that not all thoughts are memes. In principle, our immediate perceptions and emotions are not memes because they are ours alone, and we may never pass them on. We may imagine a beautiful scene from memory, or fantasise about sex or food, without using ideas that have been copied from someone else. We may even, in principle, think up a completely new way of doing something without using any memes from anyone else. However, in practice, because we use memes so much, most of our thinking is coloured by them in one way or another. Memes have become the tools with which we think.

Human thinking (indeed all thinking) may itself depend on other Darwinian processes. There have been many attempts to treat learning as a Darwinian process (e.g. Ashby 1960; Young 1965) or the brain as a 'Darwin machine' (Calvin 1987, 1996; Edelman 1989). And the idea that creativity and individual learning are selection processes is far from new (Campbell 1960; Skinner 1953). However, all these ideas concern processes entirely within one brain, while the meme is a replicator that jumps from one brain to another. Darwinian principles may apply to many aspects of brain function and development, and understanding them will be very important, but this book is just about memetics.

There are many reasons why some memes succeed and others fail. These reasons fall roughly into two categories. First, there is the nature of human beings as imitators and selectors. From the memetic point of view the human being (with its clever thinking brain) acts both as the replicating machinery, and as the selective environment for the memes. Psychology can help us understand why and how this operates. There are the properties of our sensory systems that make some memes obvious and others not, the mechanisms of attention that allow some memes to grab the available processing capacity, the nature of human memory that determines which memes will be successfully remembered, and the limitations of our capacity to imitate. We can, and will, apply this to understanding the fate of memes but it is more properly the domain of psychology and physiology than memetics.

The other kinds of reasons concern the nature of the memes themselves, the tricks they exploit, the ways they group together and the general processes of memetic evolution that favour some memes over others. These have not previously been studied by psychology and are an important aspect of memetics.

Putting together all these reasons we may be able to see why some memes succeed and others fail; why certain stories take off while others are told once and never again. Other examples include recipes, clothes fashions and interior designs; trends in architecture, rules of political correctness, or the habit of recycling glass bottles. All of these are copied from one person to another and spread by imitation. They vary slightly in the copying and some of them are more frequently copied than others. That is how we get useless popular crazes, and good ideas that never seem to get off the ground. I think there can be no doubt that memes count as replicators. This means that memetic evolution is inevitable. It is time we began to understand it.

Memes and genes are not the same

A word of caution is needed here. I have explained that the meme is a replicator and in this sense it is equivalent to the gene. However, we must not fall into the trap of thinking that memes can only work if they are like genes in other ways. This is simply not so. The science of genetics has blossomed in recent decades to the point where we can identify particular genes, map the entire human genome, and even undertake genetic engineering. Some of the insights gained from all this understanding may help us to understand memes, but alternatively some of it may just mislead us.

Also, genes are not the only other replicators to consider. For example, our immune systems are now known to work by selection. The British psychologist Henry Plotkin (1993) refers to both brains and immune systems as 'Darwin machines', and in his study of Universal Darwinism uses general evolutionary theory to apply to many other systems including the evolution of science. In each case, one can apply the ideas of replicators and vehicles (or of replicators, interactors and lineages, to use Hull's formulation) to understand the way the system evolves.

We should think of it like this – evolutionary theory describes how design is created by the competition between replicators. Genes are one example of a replicator and memes another. The general theory of evolution must apply to both of them, but the specific details of how each replicator works may be quite different.

This relationship was clearly seen by the American psychologist Donald Campbell (1960, 1965) long before the idea of memes was invented. He argued that organic evolution, creative thought and cultural evolution resemble each other and they do so because all are evolving systems where there is blind variation among the replicated units and selective retention of some variants at the expense of others. Most importantly, he explained that the analogy with cultural accumulations is not from organic evolution *per se*; but rather from a general model of evolutionary change for which organic evolution is but one instance. Durham (1991) calls this principle 'Campbell's Rule'.

We need to remember Campbell's Rule when we compare memes and genes. Genes are instructions for making proteins, stored in the cells of the body and passed on in reproduction. Their competition drives the evolution of the biological world. Memes are instructions for carrying out behaviour, stored in brains (or other objects) and passed on by imitation. Their competition drives the evolution of the mind. Both genes and memes are replicators and must obey the general principles of evolutionary theory and in that sense are the same. Beyond that they may be, and indeed are, very different – they are related only by analogy.

Some critics have tried to dismiss the whole idea of memetics on the grounds that memes are not like genes, or that the whole idea of memes is only an 'empty analogy'. We can now see why these criticisms are misguided. For example, Mary Midgley (1994) calls memes 'mythical entities' that cannot have interests of their own, 'an empty and misleading metaphor', a 'useless and essentially superstitious notion'. But Midgley has misunderstood the way in which replicators can be said to have power or 'interests of their own' and therefore she simply misses the strength and generality of evolutionary theory. Memes are no more 'mythical entities' than genes are – genes are instructions encoded in molecules of DNA – memes are instructions embedded in human brains, or in artefacts such as books, pictures, bridges or steam trains.

In a radio debate, Stephen Jay Gould (1996*b*) called the idea of memes a 'meaningless metaphor' (though I am not sure one can actually have a meaningless metaphor!). He goes even further and rejects the very notion that ideas and culture can evolve, pleading 'I do wish that the term "cultural evolution" would drop from use' (Gould 1996*a*, pp. 219–20), but I do not think that it will, because culture does evolve.

Gould seems to think that because memes and genes are related by analogy or metaphor we would somehow be doing a disservice to biological evolution by making the comparison. Again, he has missed the point that both are replicators but they need not work in the same way.

My own view is that the idea of memes is an example of the best use of analogy in science. That is, a powerful mechanism in one domain is seen to operate in a slightly different way in an entirely new domain. What begins as an analogy ends up as a powerful new explanatory principle. In this case, the most powerful idea in all of science – the explanation of biological diversity by the simple process of natural selection – becomes the explanation of mental and cultural diversity by the simple process of memetic selection. The overarching theory of evolution provides a framework for both.

With Campbell's Rule in mind we can now go on to the task of trying to understand the evolution of memes. We may use the gene as an analogy but must not expect too close a comparison. Instead, we must rely on the fundamental principles of evolutionary theory to guide us in understanding just how memes work.

Copy me!

What is special about the sentence 'Say me!' – or 'Copy me!' – or 'Repeat me!'?

They are simple (perhaps the simplest possible) examples of self-replicating sentences. Their whole point is to get themselves copied. These sentences are certainly memes – but probably not very effective ones. I doubt you will now go around shouting 'Say me!' to all your friends, but there are tricks that can be added to the simpler sentence to improve its copying potential. Hofstadter (1985) wrote about such 'viral sentences' in his monthly column in the magazine *Scientific American* called 'Metamagical Themas', and readers wrote in with many more examples.

Take: 'If you copy me, I'll grant you three wishes!' or 'Say me or I'll put a curse on you!' Neither of these is likely to be able to keep its word and few people over the age of five are likely to fall for such simple-minded threats and promises. Unless – Hofstadter adds – you simply tack on the phrase 'in the afterlife'.

In fact, it is often at about the age of five that many of us meet such sentences for the first time. I well remember being excited when I received in the post a letter that contained a list of six names and instructed me to send a postcard to the first name on the list. I was to put my own name and address at the bottom and send the new list to six more people. It promised me I would receive lots of postcards. I do not remember whether my mum prevented me from joining in or not. She might have been wise to – realising, though she would not have put it that way, that my meme–immunological system was not yet well developed. I certainly do not remember a deluge of postcards.

As these things go, that was a fairly innocuous chain letter, consisting of just a promise (the postcards) and an instruction to pass it on. At worst I would have wasted seven stamps and a postcard. I might even have received a few cards myself. Many are much more sinister, such as pyramid selling schemes, that can lose people fortunes. You would think such trivial schemes would die out, but they do not seem to. Only recently, I received an e-mail that said 'Do you like to play those scratcher lottery tickets?' (I do not) 'Would you like to learn how to turn 6 tickets into thousands?' (not particularly) 'You'll receive lottery tickets from all over the country every month! Have fun just collecting them or scratch them for the $$$jackpot$$$. There is a free service on the Web that can set you up to do just that!' Do people really join up? I suppose they must.

These are all examples of groups of memes that are replicated together. Dawkins calls such groups 'coadapted meme complexes', a phrase recently abbreviated to 'memeplexes' (Speel 1995). Memetic jargon is changing so fast and much of it is so poorly thought out and so misused that I shall try to avoid using it. However, 'memeplex' is a handy word for an important concept and so it is one of the few new words I shall adopt.

Genes, of course, go around in groups too. They clump together into chromosomes, and chromosomes are packed together inside cells. Perhaps more importantly, the whole gene pool of a species can be seen as a group of mutually cooperating genes. The reason is simple: a free-floating piece of DNA could not effectively get itself replicated. After billions of years of biological evolution, most of the DNA on the planet is very well packaged indeed, as genes inside organisms that are their survival machines. Of course, there are occasional 'jumping genes' and 'outlaw genes' and little bits of selfish DNA hitchhiking on the rest, and there are viruses that are minimal groups that exploit the replicating machinery of other larger groups – but groups, by and large, are necessary for genes to get around at all.

We could simply draw the analogy and say that memes should behave the same way but it is better to go back to the basics of evolutionary theory. Imagine two memes, one 'send a scratchcard to *x*' and another 'win lots of money'. The former instruction is unlikely to be obeyed just on its own. The latter is tempting but includes no instruction on how to. Together, and with some other suitable co-memes, the two can apparently get people to obey – and copy the whole package on again. The essence of any memeplex is that the memes inside it can replicate better as part of the group than they can on their own. We shall meet many more examples of memeplexes in due course.

The simple self-replicating meme groups we have considered so far have been given a great boost by the advent of computers and the Internet. Computer viruses are an obvious and familiar example. They can leap from user to user and the number of users (at least at the moment) keeps increasing. They can cross vast distances at the speed of light and then lie dormant in safe and solid memory banks. However, they cannot be just a bare instruction to 'Copy me'. This might succeed in clogging up the entire memory of the first computer it got into but would have no way of getting any further. So viruses have co-memes for promoting their survival. They lurk in the programs that people mail to their friends on disks. Some evade immediate detection by infecting only a small proportion of the machines they reach, and some are triggered probabilistically. Some bury themselves in memory only to pop up at a specified time – we may expect many at midnight on 31 December 1999 – quite apart from the looming problem of computers that cannot cope with the year '00'.

Some have quite funny effects, such as making all the letters on a computer screen fall to the bottom of the page – with a devastating effect on the user, but some have clogged up entire networks and destroyed books and doctoral theses. My students have recently encountered a virus on the word processor Word 6.0 that lives in a formatting section called 'Thesis' – tempting you to get infected just when your year's work is almost finished. No wonder networks are now protected by frequent automated virus checks and we have a proliferation of anti-virus software – medication for the infosphere.

Internet viruses are a relatively new arrival. I once received 'Penpal Greetings', apparently a very kind warning from someone I have never met. 'Do not download any message entitled "Penpal Greetings"' it said – and went on to warn me that if I read this terrible message I would have let in a 'Trojan Horse' virus that would destroy everything on my hard drive and then send itself on to every e-mail address in my mail box. To protect all my friends, and the worldwide computer network, I had to act fast and send the warning on to them.

Have you spotted it? The virus described does not make sense – and does not exist. The real virus is the warning. This is a very clever little memeplex that uses both threats and appeals to altruism to get you – the silly, caring victim – to pass it on. It is not the first – 'Good Times' and 'Deeyenda Maddick' used a similar trick. 'Join the Crew' is slightly more damaging, warning 'do not open or look at any mail that says RETURNED OR UNABLE TO DELIVER. This virus will attach itself to your computer components and render them useless. Immediately delete . . . there is NO remedy'. Anyone who does not spot the trick will presumably delete all those messages they had sent to people whose addresses have changed or whose e-mail systems were temporarily unavailable. A little bit of self-replicating code, using a combination of humans and computers as its replicating machinery, can have annoying consequences.

What will happen next? As people become familiar with these viruses they may learn to ignore the warnings. Thus the original type of virus will start to fail but it might let in something worse, as people start ignoring warnings they ought to heed. But then again, if ordinary old-fashioned chain letters still work, perhaps things will not change so very rapidly.

All this talk of viruses makes me wonder just why we call some pieces of computer code a virus and others a computer program. Intrinsically, they are both just lines of code, bits of information or instructions. The word is, of course, taken directly by analogy from biological viruses and probably based on the same intuitions about the way these bits of code spread. The answer is not so much to do with the harm they do — indeed some really do very little – but to do with their function. They have none apart from their own replication.

Bacteria are more complex than viruses and can be positively helpful as well as harmful. Lots live in symbiosis with us and with other animals and plants. Many do important jobs inside our bodies. Some have been co-opted to make special foods for us and so on. Viruses do little else than replicate themselves – and then only by stealing other organisms' replicating machinery. So the comparison with today's rather simple computer viruses is apt.

Might we build the equivalent of computer bacteria? Perhaps this would be a better term for some existing programs that are deliberately used to infect computer systems and run around doing jobs like updating databases or seeking out errors. Dawkins (1993) imagines useful self-replicating programs that might carry out market research by infecting many computers and then, as occasional copies got back to their starting place, providing useful statistics on user habits. Simple robotic programs, or bots, are already designed to roam around congested communications networks leaving trails that provide information about the best and worst areas of congestion, or to mimic human users in games and virtual environments. Might such simple creatures gang up together to create powerful groups just as genes have done?

These ideas seem to stretch the analogy with biological viruses a bit far (and we must be very careful of such analogies), but they do remind us that replicators vary in their usefulness. We tend to call something a virus when it is clearly acting mainly for its own replication by stealing the replicating resources of some other system – and especially when it does harm to that system. We usually give it a different name when it is useful to us.

Just the same can be seen in the world of the mind. Dawkins (1993) coined the term 'viruses of the mind' to apply to such memeplexes as religions and cults – which spread themselves through vast populations of people by using all kinds of clever copying tricks, and can have disastrous consequences for those infected. Children's games and crazes spread like infections (Marsden 1998*a*), and Dawkins suggested that children are vulnerable to 'mental infections' that more sophisticated adults can easily reject. He tried to distinguish useful memeplexes, such as science, from viral ones – an issue to which we will return.

This theme has been taken up in popular books on memetics, such as Richard Brodie's *Virus of the Mind* (1996) and Aaron Lynch's *Thought Contagion* (1996), both of which provide many examples of how memes spread through society and both of which emphasise the more dangerous and pernicious kinds of memes. We can now see that the idea of a virus is applicable in all three worlds – of biology, of computer programs and of human minds. The reason is that all three systems involve replicators and we call particularly useless and self-serving replicators 'viruses'.

But if the theory of memetics is right, viruses are not the only memes, and memetics should not become a science of mind viruses. Indeed, the vast majority of memes (like the vast majority of genes) cannot be considered as viral at all – they are the very stuff of our minds. Our memes is who we are.

According to Dennett, our minds and selves are created by the interplay of the memes. Not only are memes replicators like genes (and fit his evolutionary algorithm perfectly) but human consciousness itself is a product of memes. He has shown how the competition between memes to get into our brains has made us the kinds of creatures we are. As he puts it 'The haven all memes depend on reaching is the human mind, but a human mind is itself an artefact created when memes restructure a human brain in order to make it a better habitat for memes' (Dennett 1991, p. 207).

On this view we cannot possibly hope to understand the nature and origins of the human mind without an effective theory of memetics. But before beginning to build that theory I want to consider some previous attempts at describing the evolution of ideas. To understand the special contribution of memetics we need to understand how it differs from other theories of cultural evolution.

CHAPTER 3

The evolution of culture

From the early days of Darwinism analogies have been drawn between biological evolution and the evolution of culture. Darwin's contemporary Herbert Spencer studied the evolution of civilisations, which he viewed as progressing towards an ideal something like that of Victorian English society. Lewis Morgan's evolutionary theory of society included the three stages of savagery, barbarism and civilisation. The historian Arnold Toynbee used evolutionary ideas in identifying over thirty distinct civilisations some of which were derived from others and some of which went extinct, and even Karl Marx used evolutionary analogies in his analysis of society. Fifty years after Darwin, the American psychologist James Baldwin said that natural selection was not merely a law of biology but applied to all the sciences of life and mind, an early version of Universal Darwinism (Baldwin 1909), and he coined the term 'social heredity' to describe the way individuals learn from society by imitation and instruction (Baldwin 1896).

In some ways it is obvious that ideas and cultures evolve – that is, changes are gradual and build on what went before. Ideas spread from one place to another and from one person to another (Sperber 1990). Inventions do not spring out of nowhere but depend on previous inventions, and so on. However, truly Darwinian explanations require more than just the idea of accumulating changes over time. As we shall see, some theories of cultural evolution are little more than this idea; others try to specify a mechanism but still come back to biological evolution as the only driving force, while just a few involve the concept of a second replicator as memetics does. This is what makes memetics so distinctive and so powerful. The whole point of a memetic theory of cultural evolution is to treat memes as replicators in their own right. This means that memetic selection drives the evolution of ideas in the interests of replicating the memes, not the genes. This is the big difference that separates memetics from most previous theories of cultural evolution.

Language provides a good example of cultural evolution. Darwin pointed out the parallel between species and different languages: 'We find in distinct languages striking homologies due to community of descent, and analogies due to a similar process of formation . . . A language, like a species, when extinct, never . . . reappears' (Darwin 1859, p. 422). He also spoke of words competing for survival. Darwin probably knew about the work of the British Judge Sir William Jones who, in 1786, found remarkable similarities between Sanskrit, Greek and Latin, and concluded that all three languages must have sprung from a common source. But Darwin could not have seen many languages become extinct in his own lifetime nor have known just how many are now threatened. On a recent estimate, about 80 per cent of North American Indian languages are spoken largely by adults only, and are therefore likely to become extinct when those adults die. Similarly, about 90 per cent of Australian languages and perhaps 50 per cent of languages worldwide are doomed (Pinker 1994).

Nowadays, comparative linguists analyse the minute details of similarities and differences. They can often trace words back through many types of change such as the dropping of syllables and shifts in pronunciation. Thus, the evolutionary history of various languages can be accurately traced. Family trees of languages have been constructed that are comparable with the genetic family trees based on differences in DNA. Also, the migratory histories of whole peoples can be deduced from the languages that remain today. In Africa, for example, the 1500 or more surviving languages fall into just five main language groups, largely spoken by distinct racial groups, and their distribution can reveal which groups defeated others in the past. From a few remaining words it can be deduced that the pygmies once had their own languages but were forced into adopting those of neighbouring black farmers, and that Semitic languages, the languages of the Bible and of Islam, came not from the Near East but from Africa. The American physiologist and evolutionary biologist Jared Diamond (1997) uses language analysis as just one part of his masterful history of humanity over the past 13 000 years. He explains how languages evolve along with the people who speak them, but he does not consider the elements of language as replicators in a new evolutionary process.

In his book *The Language Instinct*, Steven Pinker (1994) explicitly applies evolutionary thinking to the development of languages, looking at heredity, variation and the effects of isolation in allowing sets of variations to accumulate. However, he does not use the idea of a selfish replicator to understand language evolution and nor does he explain why language evolved in the first place. Perhaps the answer seems too obvious – that it was biologically adaptive. But, as we shall see, this is not necessarily the right answer, and memetics can provide new twists to the argument.

Inventions as memes

Another example is the spread of inventions. Probably the most important of all 'inventions' in human history was that of farming. Although there are still many arguments over the details, archaeologists generally agree that before about 10 000 years ago all humans lived by hunting and gathering. Dating from around that time, finds in the Middle East include grains that are larger, and sheep and cattle that are smaller than their wild relatives and presumably domesticated. Farming then spread in a great wave, reaching places like Ireland and Scandinavia by about 4500 years ago. Just how many times food production arose independently is not known for sure although probably at least five times and possibly many more (Diamond 1997).

Diamond has explored the whole vexed question of why some peoples in some parts of the world ended up with all the goods – from food production to guns, germs and steel – while some have ended up still hunting and foraging, and others were completely wiped out. His answer has little to do with the innate abilities of the people themselves and everything to do with geography and climate. Food production, and the skills that went along with it, could spread easily across Europe with its East–West axis, but could not spread easily in the Americas with their North–South axis, dramatic climate variations, deserts and mountain ranges. Australia had no suitable domesticable animals, after the first humans who arrived obliterated the tame creatures they found there, and other islands, like New Guinea, are so mountainous and variable that techniques suitable in one place are unsuitable a few miles away. With this kind of analysis Diamond has explained how farming spread, bringing more complex societies in its wake.

But why did farming spread at all? The answer might seem to be obvious – for example, that farming makes life easier or happier, or that it provides a genetic advantage to the people who practise it.

In fact, it seems that farming did not make life easier, nor did it improve nutrition, or reduce disease. The British science writer Colin Tudge (1995) describes farming as 'the end of Eden'. Rather than being easier, the life of early farmers was utter misery. Early Egyptian skeletons tell a story of a terrible life. Their toes and backs are deformed by the way people had to grind corn to make bread; they show signs of rickets and of terrible abscesses in their jaws. Probably few lived beyond the age of thirty. Stories in the Old Testament describe the arduous work of farmers and, after all, Adam was thrown out of Eden and told 'In the sweat of thy face shalt thou eat bread'. By contrast modern hunter-gatherers have been estimated to spend only about fifteen hours a week hunting and have plenty of time for leisure. This is despite the fact that they have been pushed into marginal environments far poorer than those in which our ancient ancestors probably lived. Why would people the world over have given up an easier life in favour of a life of toil and drudgery?

Tudge assumes 'that agriculture arose because it was favoured by natural selection' (1995, p. 274) and therefore looks for a genetic advantage. He suggests that because farming produces more food from a given area of land, farmers will produce more children who will encroach on neighbouring hunter-gatherer's lands and so destroy their way of life. For this reason, once farming arrives no one has the luxury of saying 'I want to keep the old way of life'. However, we know from the skeletons of early farmers that they were malnourished and sickly. So was there really a genetic advantage?

Memetics allows us to ask a different question. That is, why were farming practices successful as memes? In other words, how did these particular memes get themselves copied? The answers might include their benefits to human happiness or to human genes, but are not confined to those possibilities. Memes can spread for other reasons too, including less benign ones. They might spread because they *appear* to provide advantages even when they do not, because they are especially easily imitated by human brains, because they change the selective environment to the detriment of competing memes, and so on. With a meme's eye view we ask not how inventions benefit human happiness or human genes, but how they benefit themselves.

Turning to more modern technology, from the invention of the wheel to the design of cars, there is plenty of evidence that innovations evolve in the sense that they arise from what went before. In *The Evolution of Technology*, George Basalla (1988) develops an evolutionary account of the way in which hammers, steam engines, trucks and transistors have come about. Playing down the importance of heroic inventors he emphasises the gradual process of change through imitation and variation. For example, many features of wooden buildings were reproduced in stone by the Greeks, the first iron bridge built in the late 1770s was modelled on woodworking practices, and even the humble plastic bucket often still shows signs of its origins in metal. Transistors were only gradually miniaturised and radio signals very gradually transmitted further and further.

Basalla questions the idea of technology making progress towards any grand goal such as 'the advancement of humanity' or 'the overall betterment of the human race' (Basalla 1988). In true Darwinian fashion he sees technology as developing only from the present situation with very limited specific goals and suggests we discard the entire illusion of technological progress. But I would add here another word of caution concerning the word 'progress'. The word can be used in at least two different ways. One implies progress towards some goal or objective; the other implies only increasing design, increasing complexity, or any kind of continuous development *without* a particular goal or end point built in. Basalla, like Gould, throws out both kinds of progress. I would throw out only the first. Today's technology is far more sophisticated and complex than that of 10 000 years ago, and that is progress of the second kind. But, there is no progress towards some predetermined or ultimate goal. We did not have to go from stone axes to fax machines – we did have to go from stone axes to something more specialised, more designed and more improbable. In Dennett's terminology, there has been ever more exploration of the Design Space of possible artefacts. In Dawkins's terminology, technology has been slowly climbing its own Mount Improbable. This is technological progress, if not progress towards anything in particular.

So why do we have fax machines? Why Coca Cola cans and wheelybins? Why Windows 98 and felt-tip pens? I want answers to these specific questions. 'Because we want them' is not a sufficient answer. 'Because we need them' is clearly untrue. If we want to understand how the fantastic complexity of our technological world came about it is not enough just to say that technology evolves, without providing a mechanism. In later chapters I shall explain how a memetic approach can help.

Scientific ideas also evolve and there have been many theories that attempt to explain how. The influential philosopher Karl Popper, in one of his best known contributions to the philosophy of science, suggested that scientific knowledge is gained by the falsification of hypotheses, not by accumulating proof or evidence *for* theories. Science can then be seen as a competitive struggle between rival hypotheses in which only some survive.

Popper also applied Darwinian thinking in his three 'cosmic evolutionary stages': World 1 is the world of physical objects such as trees, tables and human bodies; World 2 is the world of subjective experiences including feelings, emotions and consciousness; and World 3 is the world of ideas; of language and stories, works of art and technology, mathematics and science. World 3 is largely autonomous, even though created by us (Popper 1972), and its contents have effects on the other worlds by a kind of downward causation. So, for example, scientific theories may appear as World 1 objects (the scientist, the journal papers, the experimental apparatus, and so on), but they are more than just physical objects. The ideas *themselves* influence those objects. The problems, hypotheses, theories and intellectual struggles work through World 2 and into World 1. Scientific ideas really do change the world: 'once theories exist, they begin to have a life of their own' (Popper and Eccles 1977, p. 40).

How can an idea change the physical world? Popper was struggling here with a difficult and important problem, related to the value of reductionism in science and the viability of materialism as a world view. I do not think he solved it. His three worlds contain very different kinds of material and he has to propose a tricky kind of interactionism to link them. Interestingly, he touches on the role of imitation but without realising how it might help. For example, in explaining how artistic ideas can have real effects, he says 'a sculptor may, by producing a new work, encourage other sculptors to copy it, or to produce similar sculptures' (Popper and Eccles 1977, p. 39). In his terms, the ideas in the sculptor's mind (World 3) affect the experiences of others (World 2) and thus lead to new sculptures (World 1).

In memetic terms, all that happens – whether in science or art – is selective imitation. The emotions, the intellectual struggles, the subjective experiences – these are all parts of the complex system that leads to some behaviours being imitated and others not. And it is because imitation lets loose a second replicator that ideas begin to 'have a life of their own'. In this way, memetics provides a mechanism for the evolution of scientific ideas that Popper's three worlds cannot.

Although Popper did not use the idea of a replicator, his views directly gave rise to the new field of evolutionary epistemology, which does. Evolutionary epistemology began in 1974 with a critique of Popper by Campbell, and applies Darwinian thinking to the evolution of knowledge (Hull 1988*a*,*b*; Plotkin 1982). The American philosopher David Hull studies the way scientific ideas develop over time in lineages rather as species do. He treats scientific ideas as the replicators and scientists as the interactors (he prefers the term 'interactor' to Dawkins's 'vehicle' because of its more active connotations). Plotkin considers science as not only 'the product of a "Darwin machine"' but 'a special form of culture that is transformed in time by evolutionary processes' (Plotkin 1993, pp. 69, 223). According to evolutionary epistemology, biological adaptations are one form of knowledge, and science is another; both are produced by the processes of blind variation and selective retention (Campbell 1975). This approach is firmly based in Universal Darwinism and does not bring everything back to genetic advantage.

Whose advantage?

We can now see that many theories of cultural change use evolutionary ideas but they are not the same as memetics. There are two fundamental differences. First, most do not distinguish general evolutionary theory from the specifics of biological evolution. This means they are unclear about the relationship between biology and culture and easily fall foul of the obvious differences between genetics and cultural evolution. Second, they do not introduce the idea of a second replicator such as the meme. This means they do not see cultural evolution as proceeding in the interests of a selfish replicator.

This last issue is most important and I want to pursue it. The whole point of memetics is to treat the meme as a replicator in its own right, operating entirely for the benefit of its own selfish replication. If there is no second replicator, and you are a committed Darwinian, then somehow or other everything must come back to the genes – to biological advantage. If there are two replicators (or more) then there will inevitably be conflicts of interest – circumstances in which the interests of the genes pull in one direction and those of the memes in the opposite direction. These examples are very important for memetics because they would not be predicted by a purely genetic theory. If they occur, they prove that we need a theory of memes – or at least a theory involving some kind of second replicator. This is what distinguishes memetic theory from other theories of cultural evolution.

Dennett (1995) makes the same point when he asks '*Cui bono*?', who benefits? He says 'The first rule of memes, as it is for genes, is that replication is not necessarily for the good of anything; replicators flourish that are good at . . . replicating! . . . The important point is that there is no *necessary* connection between a meme's replicative power, its "fitness" from *its* point of view, and its contribution to *our* fitness (by whatever standard we judge that)' (Dennett 1991, p. 203, italics in the original).

Dawkins explains:

> As soon as the primeval soup provided conditions in which molecules could make copies of themselves, the replicators themselves took over. For more than three thousand million years, DNA has been the only replicator worth talking about in the world. But it does not necessarily hold these monopoly rights for all time. Whenever conditions arise in which a new kind of replicator *can* make copies of itself, the new replicators *will* tend to take over, and start a new kind of evolution of their own. Once this new

> evolution begins, it will in no necessary sense be subservient to the old (Dawkins 1976, pp. 193–4, italics in the original).

Of course, memes could only come into existence when the genes had provided brains that were capable of imitation – and the nature of those brains must have influenced which memes took hold and which did not. However, once memes had come into existence they would be expected to take on a life of their own.

Dawkins argued that biologists had so deeply assimilated the idea of genetic evolution that they tended to forget that it is only one of many possible kinds of evolution. He complained of his colleagues that 'In the last analysis they wish always to go back to "biological advantage"' (Dawkins 1976, p. 193). In other words, they might accept the idea of memes, or some kind of unit of cultural evolution, but then still believe that memes must always act somehow for the benefit of the genes. But this is missing the whole point of the second replicator. If memes are replicators, as I am convinced they are, then they will not act for the benefit of the species, for the benefit of the individual, for the benefit of the genes, or indeed for the benefit of anything but themselves. That is what it means to be a replicator.

I am labouring this point because I am now going to review some theories of cultural evolution that *have* introduced the idea of a second replicator – or at least some kind of new cultural unit. (Durham 1991 provides a more thorough review.) At first sight these may all appear equivalent to the idea of the meme, but they are not. There are many similarities and differences but the most important point to look for is whether the new unit is really being treated as a replicator in its own right. If it is not then the theory is not equivalent to memetics.

In 1975, just before Dawkins proposed the idea of memes, the American anthropologist F. T. Cloak wrote about cultural instructions. He pointed out that whenever we see any behaviour being performed we assume that there is some internal structure in the animal's nervous system that causes that behaviour. All animals have such instructions but humans, unlike other animals, can acquire new instructions by observing and imitating others. Cloak suggested that culture is acquired in tiny, unrelated snippets that he called 'corpuscles of culture' or 'cultural instructions'.

Furthermore, he distinguished very carefully between the instructions in people's heads and the behaviour, technology or social organisation that those instructions produce. The former he called the 'i-culture' and the latter the 'm-culture'.

He was absolutely clear about the status of cultural instructions, even though he did not use the replicator concept. He said that the ultimate function of both i-culture and m-culture is the maintenance and propagation of the i-culture. Therefore, he concluded, we should not be surprised to find some m-culture features that perform functions that are irrelevant, or even destructive, to the organisms who make or do them. He compared cultural instructions to parasites that control some of their host's behaviour – a bit like a flu virus that makes you sneeze to get itself propagated. He concluded 'In short, "our" cultural instructions don't work for us organisms; we work for them. At best, we are in symbiosis with them, as we are with our genes. At worst, we are their slaves' (Cloak 1975, p. 172). Quite clearly, Cloak had seen the implications of having a second selfish replicator – even though others subsequently argued that cultural instructions are not replicators at all (Alexander 1979).

In *The Selfish Gene*, Dawkins mentions Cloak, saying that he wants to go further in directions being explored by Cloak and others. However, Dawkins lumps together both the behaviours and the instructions that produce them, and calls them all memes, while Cloak separates the two – a distinction that is somewhat analogous to the distinction between the genotype and the phenotype in biology. Later, Dawkins (1982) makes the same distinction as Cloak and defines a meme as 'a unit of information residing in a brain'. I shall return to consider the importance of this difference later on. For now we need only note that Cloak's cultural instruction is, like the meme, a true second replicator.

Sociobiology and culture on a leash

While Dawkins was writing *The Selfish Gene*, the new science of sociobiology was being established – studying the genetic and evolutionary basis of behaviour. There was, at the time, a great outcry against applying sociobiology to human behaviour. Some of this came from sociologists, anthropologists and others who argued that human behaviour was almost entirely free from the constraints of the genes and could not be understood by what they saw as (horror of horrors) 'genetic determinism'. The genes, they claimed, only give us a 'capacity for culture'. Some came from ordinary people who rejected the idea that their cherished beliefs, decisions and actions were constrained by their genetic make-up – what about 'free will'?

This reaction reminds me of the antagonism to Newton, to Copernicus and to Darwin himself. Sociobiology seemed to push human beings further off their self-created pedestal – to undermine their sense of free will and autonomy. As we shall see, memetics takes a further big step in this direction and so will probably reap the same antagonism. Still – as Cloak put it '. . . if we *are* the slaves of some of "our" cultural traits, isn't it time we knew it?' (Cloak 1975, p. 178).

Much of the antagonism to sociobiology has died down, perhaps because of the increasing evidence for the evolutionary basis of human behaviour, and perhaps because of a better understanding of the way genes and environment interact. The old image of genes as providing a blueprint or wiring diagram for building a body is clearly wrong. A better analogy is with a recipe, though it is still not a close one. Genes are instructions for building proteins, and the results of this protein synthesis are influenced at every stage by the available raw materials and the nature of the environment. Nothing is purely genetically *determined* and nothing purely environmentally *determined*. We human beings, like all other creatures, are a complex product of both – and this is true of the way we behave as well as the shape of our legs.

In spite of the antagonism, sociobiology made great progress but, as its founding father Edward O. Wilson complained, it had little to say about the individual human mind or the diversity of cultures. In 1981, Wilson teamed up with the physicist Charles Lumsden to develop a theory of gene–culture coevolution and introduced the concept of the 'culturgen' as 'the basic unit of inheritance in cultural evolution' (Lumsden and Wilson 1981, p. x). They hoped their new theory would lead right through from genes to mind to culture and developed mathematical treatments of how different culturgens would affect genetic fitness. However, they always came back to the genes as the final arbiters. If maladaptive culturgens are sometimes selected this is because their harm is not immediately apparent and so there is some lag before the system adapts. Ultimately, the genes will win out. As they put it – 'the genes hold culture on a leash'.

The 'leash principle' is a more memorable way of expressing what Dawkins meant about his colleagues wanting 'always to go back to "biological advantage"'. It also provides us with a helpful image. If Lumsden and Wilson are right then the genes are always the owner and the culturgens are the dog. The leash can sometimes get longer – even extremely long – but it is still a dog at the other end. According to memetics, the genes may turn into a dog and the memes become the owner – or perhaps we should enjoy the spectacle of two dogs, one on either end – each running like mad to serve their own selfish replication.

The Stanford geneticists, Luigi Cavalli-Sforza and Marcus Feldman (1981), developed a detailed model of cultural transmission based on the 'cultural trait' as the unit. Cultural traits are learned by imprinting, conditioning, observation, imitation or direct teaching (note that this is a broader range than for memes which, by definition, have to be passed on by imitation and cannot be acquired by imprinting or conditioning). They clearly distinguish cultural selection from Darwinian or natural selection and they use the concept of 'cultural fitness' – that is, the fitness for survival of a cultural trait itself – a concept that is useful in memetics. They also introduced the distinction between vertical transmission – such as from parent to child – and horizontal transmission – such as from child to child or adult to unrelated adult. We shall see later how important this is for understanding life in an age of predominantly horizontal transmission.

Cavalli-Sforza and Feldman listed different mechanisms of cultural transmission and provided mathematical models of particular cases, including maladaptive ones. A seriously maladaptive example is the practice of cannibalism in the funeral rites of a New Guinea highland tribe called the Foré. As part of complex rituals honouring their dead the Foré ate parts of the human bodies. In fact, they preferred eating pork to human flesh and so the men tended to get more of this prized food, leaving the women and children to more cannibalism (Durham 1991). This practice led directly to an epidemic of the degenerative disease kuru, which killed about 2500 Foré people, mostly women and children. Cavalli-Sforza and Feldman demonstrated mathematically that a maladaptive trait like this could eliminate up to 50 per cent of its carriers and still spread through a population.

However, despite contributing so much to our understanding of cultural transmission and the spread of maladaptive practices. Cavalli-Sforza and Feldman still see 'cultural activity as an extension of Darwinian fitness' (1981, p. 362), and this is what distinguishes their theory from memetics. As Dennett (1997) puts it, they do not ask the all-important *Cui bono?* question. Or, if they do, they simply assume that the answer must be the genes, and do not consider the possibility that 'it is the cultural items *themselves* that benefit from the adaptations they exhibit' (Dennett 1997, p. 7). For Cavalli-Sforza and Feldman, cultural adaptation means the use of skills, beliefs, and so on, to the ultimate benefit of the genes – and the term 'maladaptive' means maladaptive to the genes. Even if only in the long run, they say, 'The mechanism of natural selection retains ultimate control' (Cavalli-Sforza and Feldman 1981, p. 364). In other words they too believe in the leash.

The only anthropologists who seem to have let go of the leash are Robert Boyd and Peter Richerson from the University of California at Los Angeles. Like sociobiologists, they accept that culture arises from 'natural origins' but claim that models that take cultural evolution into account – like their 'Dual Inheritance Model' – can do better than sociobiology. They refer to Campbell's rule and are convinced, as I am, that cultural variants must be subject to their own form of natural selection. They analyse in great detail the structural differences between cultural transmission and genetic transmission and conclude '. . . the behaviour that enables an individual to maximize his chance to enculturate cultural offspring may not be the behavior that will maximize the transmission of genes to the next generation' (Boyd and Richerson 1985, p. 11). In their version of coevolution the genes can keep culture on a leash, culture can keep the genes on a leash, or the two may evolve in competition or mutuality (Richerson and Boyd 1989). They seem to be truly treating their cultural unit as a separate replicator. Boyd and Richerson are anthropologists, concerned far more than I shall be with cultural variation. However, many of their ideas will prove useful in understanding the selection of memes.

The anthropologist William Durham uses the term 'meme' for his unit of cultural evolution, and at first sight may appear to take a memetic view, but a closer look shows that for him the meme is not truly a selfish replicator. He claims that organic and cultural selection work on the *same* criterion – that is, inclusive fitness – and are complementary. He argues that Boyd and Richerson take 'the abstract genetic analogy a bit too far' and are 'strongly anti-Darwinian', and he does not agree with them that human evolution is *fundamentally different* from that of other organisms (Durham 1991, p. 183).

This comes to the heart of the issue. For me, as for Dawkins and Dennett, memetic evolution means that people *are* different. Their ability to imitate creates a second replicator that acts in its own interests and can produce behaviour that is memetically adaptive but biologically maladaptive. This is not just a temporary aberration to be ultimately reined in by the powerful genes, but is permanent, because memes are powerful in just the same way that genes are; they have replicator power. Cloak, and Boyd and Richerson, seem to agree but the others do not accept the independent replicator power of their units of cultural transmission. In that important sense they are much closer to traditional sociobiology – their motto might be 'the genes will always win'. The leash may sometimes get very long but the dog can never get away.

That brings us full circle to the modern successor to sociobiology, which largely takes the same standpoint. Evolutionary psychology is based on the idea that the human mind evolved to solve the problems of a hunter-gatherer way of life in the Pleistocene age (Barkow *et al.* 1992; Pinker 1997). In other words, all our behaviours, beliefs, tendencies and customs are adaptations. For example, sexual jealousy and love for our children, the way we acquire grammar or adjust our food intake to deal with nutritional deficits, our avoidance of snakes and our ability to maintain friendships are all seen as adaptations to a lifestyle of hunting and gathering. Evolutionary psychologists therefore argue that all behaviour ultimately comes back to biological advantage.

Evolutionary psychology can take us a long way, but is it far enough? I say not. From the perspective of memetics, evolutionary psychology provides a crucial underpinning. In order to understand why certain memes are positively selected and others rejected we need to understand the way natural selection has moulded our brains for the benefit of the genes. We like sweet cakes and caffeine-filled drinks, we look twice at a magazine with a naked woman on the front and not at the one with trains on it. We buy bright bunches of flowers and avoid the smell of rotting cabbages, and all this is essential to understanding memetic selection. But it is not the whole story. To fully understand human behaviour we must consider both genetic *and memetic* selection. Most evolutionary psychologists reject outright the idea that a second replicator is needed. My task in this book must be to show why it is.

....

I have explored various approaches to cultural evolution to see whether any use the same ideas as memetics but under another name. The answer, with the limited exceptions I have discussed, is no. It seems that there is no ready-made science of memetics waiting to be taken over. If we need a science of memetics, as I am convinced we do, then we shall have to build one up from scratch.

The main tools available are the basic principles of evolutionary theory, the founding ideas of Dawkins, Dennett and other early memeticists, and the relevant ideas from a cultural anthropology discussed above. Of course, we can also draw on more than a century of research in psychology and several decades of cognitive science and neuroscience.

Using these tools, I shall try to lay the foundations for a science of memetics. I can then use this to provide new answers to old questions from the apparently trivial, such as 'Why is my head so full of thoughts?' to the weighty question of why human beings have such big brains. The first step in this endeavour is to start looking at the world from a meme's eye view.

CHAPTER 4

Taking the meme's eye view

We can now start to look at the world in a new way. I shall call this the meme's eye view, though, of course, memes do not really have eyes or points of view. They cannot see anything and they cannot predict anything. However, the point of this perspective is the same as the 'gene's eye view' in biology. Memes are replicators and tend to increase in number whenever they have the chance. So the meme's eye view is the view that looks at the world in terms of opportunities for replication – what will help a meme to make more copies of itself and what will prevent it?

I like to ask a simple question – indeed I shall use this question again in several different contexts. Imagine a world full of hosts for memes (e.g. brains) and far more memes than can possibly find homes. Now ask, which memes are more likely to find a safe home and get passed on again?

This is a reasonable way to characterise the real world we live in. Each of us creates or comes cross countless memes every day. Most of our thoughts are potentially memes but if they do not get spoken they die out straight away. We produce memes every time we speak, but most of these are quickly snuffed out in their travels. Other memes are carried on radio and television, in written words, in other people's actions, or the products of technology, films and pictures.

Think for a moment about all the thoughts you have had in the past ten minutes – let alone all day. Even while reading you have probably thought about other people, remembered things you meant to do, made plans for later in the day, or (I hope) pursued ideas sparked off by the book. Most of these thoughts will never be thought again. You will not pass them on and they will perish.

Think of the number of things you are likely to say to someone else today – or the number of words you will hear other people speak. You might listen to the radio, watch television, have dinner with other people, help your children with the homework, answer the phone to people far away. Most of what is said in these conversations will never be passed on again. Most of it will *not* reappear as 'Then he said to her . . .' or 'And did you know . . .'. Most will die at birth.

Written words may not fare much better. The words on this page have at least got as far as being read by you, but may well get no further. Even if you do pass them on, you may scramble them for easier recall or because I have not made myself clear, so the copying fidelity will not always be high. Millions of newspapers are printed each day but by a week later most of the copies have gone and most people have forgotten what was in them. Books may do a little better – though in the United States alone something like one hundred thousand new books are published every year. Not all of them can be influential or memorable. And while some scientific papers are widely read and quoted, it is rumoured that the majority are not read by anyone at all!

We could not (even in principle) calculate the proportion of potential memes that actually do get passed on but the idea is clear enough. There is enormous selection pressure, and therefore very few survivors from among the very many starters. Only a few memes are successfully copied from brain to brain, from brain to print, from print to print or from voice to compact disk. The ones we regularly meet are the successful ones – the ones that made it in the competition for replication. My question is simply – which memes are those?

I am going to take the meme's eye view as a way of tackling several controversial questions. I shall start with a simple one. The question itself may not be profound but it does turn out to be rather intriguing – and it will give us practise in thinking from the meme's point of view.

Why can't we stop thinking?

Can you stop thinking?

Perhaps you have practised meditation or some other method of calming the mind. If so you will know that the task is not trivial. If you have not, I suggest you try now to empty your mind for a minute or so (or if you cannot face it now, try it sometime when you have nothing 'better' to do, waiting for the kettle to boil, or the computer to boot up, for example). When any thought comes along, as it certainly will, just acknowledge it and let it go. Do not get tangled up in the thoughts or follow them up. See whether you can find any space between them. The simplest forms of meditation are no more than this kind of practice. It is fiendishly difficult.

Why? You will doubtless notice that thoughts just seem to pop up out of nowhere and grab your attention. You may also notice what kinds of thought they are. Typically, they are imagined conversations or arguments, reruns of events with new endings, self-justifications, complicated plans for the future, or difficult decisions that have to be made. They are rarely simple images, perceptions or feelings (which can come and go without causing trouble); rather, they use words, arguments, and ideas you have acquired from other people. In other words, these incessant thoughts are memes. 'You' cannot command them to cease. You cannot even command them to go slower nor tell yourself not to get sucked into them. They seem to have a life and power of their own. Why?

From the biological point of view this constant thinking does not appear to be justified. I say this cautiously, in the recognition that many things that at first did not appear to be in the interests of the genes subsequently have turned out to be. Nevertheless it may be helpful to think this through.

Thinking requires energy. One of the many benefits of techniques like PET scanning (positron emission tomography) is that we can observe graphically what goes on in a brain when someone is thinking. Scans, although still severely limited in resolution, can show the relative amounts of blood flowing in different areas of the brain. For example, when someone is doing a visual task there is more activity in the visual cortex, when listening to music more in the auditory cortex, and so on. As had long been suspected, imagining something uses similar parts of the brain as actually seeing or hearing the same thing. So imagining conversations activates speech areas, and so on. Experiments comparing simple visual tasks with more difficult ones show higher levels of activity with the more difficult task.

The amounts of energy used are small compared with, say, running up a hill, but they are not entirely negligible. Blood flow means that oxygen and stored energy are being burned up, and these have to be worked for. If an organism could get by without thinking all the time it would use less energy and hence ought to have a survival advantage.

Presumably, then, all this thinking has some function. But what? Perhaps we are practising useful skills, or solving problems, or thinking through social exchanges so as to make better deals, or planning future activities. I have to say this does not seem to be plausible for the sorts of daft and pointless thoughts I tend to think about. However, applying evolutionary thinking to today's situation may not be appropriate. We did not evolve along with books, telephones and cities.

Evolutionary psychologists would suggest that instead we consider our hunter-gatherer past. Speculating in too much detail is dangerous since we have rather little information about the far past, but many authors have provided good descriptions based on the available evidence (Dunbar

1996; Leakey 1994; Mithen 1996; Tudge 1995). They tend to agree that people lived in groups of roughly 100–250 people, with strong family ties and complex social rules. Women tended to gather the plant foods and men to hunt. Life expectancy was short compared with today. Density of population was limited by the large area of land needed for this lifestyle and there were predators and disease to worry about. However, providing the food would not take all day and there would have been many hours left over.

In such a situation would it make sense to keep on thinking all the time? Would those endless thoughts have justified their energy costs in terms of survival advantage? Or would it have been better to save the energy and be able just to sit and not think – as cats appear to do when resting in the sun? I am only speculating but I would suggest that it may have benefited the genes more if we could stop thinking sometimes and conserve valuable resources. Why, then, can we not?

The answer from memetics is to start thinking in terms of replicators trying to get copied.

First, let us think about brains without memes. If the brain really is a Darwin machine then the thoughts, perceptions, ideas, memories, and so on, that go on inside it must all be competing for the brain's limited processing resources. Natural selection will have ensured that the brain's attention mechanisms generally devote most resources to the processing that helps the genes that made it. Within those constraints, all the thoughts and ideas will compete for attention and the chance to get copied. However, they are limited to one brain and subject to the pressures of natural selection.

Now imagine a brain capable of imitation – a brain with memes. A brain with memes not only has much more information to store, but the memes themselves are tools for thinking with (Dennett 1991). Far more kinds of thinking are possible when you have learned words, stories, the structure of arguments, or new ways of thinking about love, logic or science. There are now far more thoughts competing for the same limited processing capacity of the brain. Not only that, but memes can also get copied from one brain to another.

If a meme can get itself successfully copied it will. One way to do so is to command the resources of someone's brain and make them keep on rehearsing it, so giving that meme a competitive edge over memes that do not get rehearsed. Memes like this are not only more likely to be remembered but also to be 'on your mind' when you next speak to someone else. If we take stories as an example, a story that has great emotional impact, or for any other reason has the effect that you just cannot stop thinking about it, will go round and round in your head. This will consolidate the memory for that story and will also mean that, since you are thinking about it a lot, you are more likely to pass it on to someone else, who may be similarly affected.

We may now ask the question I posed at the start. *Imagine a world full of brains, and far more memes than can possibly find homes. Which memes are more likely to find a safe home and get passed on again?*

Compare a meme that not only grabs the attention but tends to make its host keep on mentally rehearsing it, with one that buries itself quietly in memory and is never rehearsed, or a thought that is too boring ever to think again.

Which will do better? Other things being equal, the first type will. So these are the thoughts that get passed on again while the others simply fade away. The consequence is that the world of memes – the meme pool – fills up with the kinds of thoughts that people tend to think about. We all come across them and so we all think an awful lot. The reason 'I' cannot compel myself to stop thinking is that millions of memes are competing for the space in 'my' brain.

Note that this is just a general principle designed to show why we think so much. We should also be able to find out which kinds of memes these successful ones are. For example, they may be ones that trigger certain emotional responses, or which relate to the core needs for sex and food – and evolutionary psychology can help us here. They may be ones that provide especially good tools for creating more memes, or which fit neatly into already installed memeplexes like political ideologies or belief in astrology. But exploring these reasons is a more specific task and I shall return to it later. For the moment I want only to show how general principles of memetics can help us understand the nature of our minds.

I think of this as the 'weed theory' of memes. An empty mind is a bit like my vegetable garden when I have dug and cleared and hoed it. The earth is brown, rough, rich and ready for anything that wants to grow. A week or two later there are little bits of green poking up in places; another week or two later there are serious plants dotted about; and soon the whole plot is covered in green, tangled with creepers, thrusting with tall leaves, and not a spot of brown earth can be seen. The reason is obvious. If something can grow it will. There are far more seeds in the soil and in the air than can possibly grow into mature plants, and as soon as any one of them finds itself with space, water and light, off it goes. That is just what seeds do. Memes do just the same with brains. Whenever there is any spare thinking capacity memes will come along and use it up. Even when we are already thinking about something absolutely gripping any other idea that is even more gripping may displace the first from its position, improve its chance of getting passed on, and so increase the likelihood of someone else being infected with it. On this view the practice of meditation is a kind of mental weeding.

There are other analogies in the world of biology (although we must remember they are only analogies). Take a forest, for example. In a forest every tree has to compete for light, so genes for growing tall trunks will do well and tend to spread in the gene pool, as all the trees carrying genes for shorter trunks die out in the gloom below. In the end the forest will consist of trees that all have the tallest trunks they can manage to create.

Who benefits? Not the trees. They have all invested enormous amounts of energy into growing the trunks and are still competing with each other. There is no way that they could come to a gentleman's agreement not to bother with trunks, for if some of them did, a cheat could always succeed by breaking the pact. So forests are a common creation all over the planet. The beneficiary is the successful gene, not the trees.

Returning to our poor overactive brains, we can ask again – who benefits? The constant thinking does not apparently benefit our genes, and nor does it make us happy. The point is that once memes have appeared the pressure to keep thinking all the time is inevitable. With all this competition going on the main casualty is a peaceful mind.

Of course, neither the genes nor the memes care about that – they are just mindlessly replicating. They have no foresight and they could not plan according to the consequences of their actions – even if they did care. We should not expect them to have created a happy and relaxing life for us and indeed they have not.

I have used this simple example to show the way in which I want to use memetics to understand the human mind. Later I will use the same approach to ask a closely related question – why do people talk so much? You may already think the answer is obvious, but before we explore the many ramifications of this one I want to add an important word of caution.

Not everything is a meme!

Not everything is a meme

Once you grasp the basic idea of memes it is all too easy to get carried away with enthusiasm and to think of everything as a meme – to equate memes with ideas, or thoughts, or beliefs, or the contents of consciousness, or anything you can think of. This tendency is deeply confusing and gets in the way of understanding what memes can and cannot do. We need to start with a clear and precise definition of the meme and decide just what does and does not count.

The most important point to remember is that, as in Dawkins's original formulation, memes are passed on by imitation. I have described them as 'instructions for carrying out behaviour, stored in brains (or other objects) and passed on by imitation'. The new *Oxford English Dictionary* gives **meme** (mi:m), *n. Biol.* (shortened from *mimeme* . . . that which is imitated, after GENE *n.*) An element of a culture that may be considered to be passed on by non-genetic means, esp. imitation'. Imitation is a kind of replication, or copying, and that is what makes the meme a replicator and gives it its replicator power. You could even say that 'a meme is whatever it is that is passed on by imitation' – if it didn't sound so awkward.

We may (and will) argue about just what counts as imitation but for now I shall use the word 'in the broad sense', as Dawkins did. When I say 'imitation' I mean to include passing on information by using language, reading, and instruction, as well as other complex skills and behaviours. Imitation includes any kind of copying of ideas and behaviours from one person to another. So when you hear a story and pass on the gist to someone else, you have copied a meme. The important point is that the emphasis on imitation allows us to rule out all kinds of things which cannot be passed on and therefore cannot be counted as memes.

Look away from this page for a moment and rest your eyes on the window, the wall, a piece of furniture or a plant. Anything will do, but just look quietly at it for – say – five seconds before you come back to reading. I presume you experienced something. There were sights, sounds, and impressions that made up your experience in those few seconds. Did they involve memes? Perhaps you said to yourself 'That plant needs watering' or 'I wish there weren't so much traffic outside'. If so, you were using words; you obtained those words memetically and you could pass them on again – but as for the perceptual experience itself – that does not necessarily involve memes.

Of course, you could argue that now we have language everything we experience is coloured by our memes. So let us consider the experiences of some other animal that does not have language. One of my cats will do as an example. She is not the brainiest of creatures but she does have a rich and interesting life and many capabilities despite having acquired next to nothing by imitation.

First of all she can see and hear. She can run after butterflies and scamper up a tree – which requires complex perceptual and motor skills. She can taste and smell, and choose Whiskas over Katkins. She has a powerful sense of hierarchy and territory and will hiss at or run away from some cats, and play with others. She can obviously recognise individual cats and also some humans, responding to their voices, footsteps or touch, and can communicate with them using movement, physical contact and her own quite powerful voice. Her mental map is complex and detailed. I have no idea how far it stretches but it covers at least four human gardens, two roads and many human-made and cat-made paths. She can relate the position of a person at a window to the room they are in, and find the most direct route to the kitchen when the knife hits the bowl. And when she arrives, at the word 'Hup' she neatly stands on her hind legs and tucks in her front paws.

Her life includes many of the experiences that I can recognise in my life too – perception, memory, learning, exploration, food preferences, communication and social relationships. These are all examples of experiences and behaviours that have not been acquired by imitation and so are not memes. Note that my cat has done a lot of learning in her lifetime, and some of it from me, but it cannot be 'passed on by imitation'.

If we are to be sure what is meant by a meme then we must carefully distinguish learning by imitation from other kinds of learning. Psychology traditionally deals with two major types of individual learning (i.e. learning by an individual animal or person) – classical conditioning and operant conditioning. In classical conditioning, originally studied by Pavlov with his salivating dogs, two stimuli become associated by repeated pairing. My cat has probably learned to associate certain sounds with food-time, the sight of certain cats with fear, the sound of rain with 'not a nice day to go out', and so on. Just as I have learned to freeze at the sound of a dentist's drill (and I still do, even though I have been given anaesthetics for the past 25 years!), and to relax with pleasure at the sound of the ice going in the gin and tonic. You could say that in classical conditioning some aspect of the environment has been copied into a brain, but it stops with that brain and cannot be passed on by imitation.

Operant conditioning is when a behaviour made by an animal is either rewarded or punished and therefore either increases or decreases in frequency. Skinner famously studied this kind of trial and error learning with his rats or pigeons in cages, pressing levers to obtain food. My cat probably learned to use the cat door by operant conditioning, as well as better ways of catching voles. She also learned to beg that way. At first she made feeble attempts to get her nose up to where I was holding the dish.

Then, in a process called shaping, I progressively rewarded her for ever neater and neater begging, finally hiding the dish behind my back and saying 'Hup'. And in case you think this is unfair treatment of a small weak animal by a large and powerful one I should point out that she has successfully trained me to leave my desk to come and stroke her when required.

Skinner also pointed out the similarity between operant conditioning and natural selection – some behaviours are positively selected and others weeded out. In this way learning can be seen as an evolutionary system in which the instructions for carrying out behaviour are the replicators. Several selection theories of learning and of brain development have been proposed but as long as the behaviours cannot be passed on to someone else by imitation then they do not become memes and the selection is not memetic selection.

Much of human learning is Skinnerian and not memetic. Whether consciously or not, parents shape their children's behaviour by the way they reinforce them. The best reward for children is attention, and rewards work better than punishment. So if parents pay lots of attention to their children when they are behaving well, and act uninterested when they scream or have tantrums, then behaving well is simply in the best interests of the kids and they will do it. Parents who do everything for their children end up with dependent children, while those who expect their children to find their own games kit, and leave them to reap teacher's wrath if they are late for school, end up with children who take responsibility for themselves. You may think you taught your daughter to ride a bike but in all probability you just bought the bike, provided encouragement, and trial and error did the rest. There is not necessarily anything memetic in all of this (apart from the idea of riding a bike at all). Much of what we learn, we learn only for ourselves and cannot pass on.

In practice, we can probably never tease out those things we have personally learned by imitation from those we have learned in other ways – but in principle the two are different. We know lots of things that are not memes. Some authors, however, imply that virtually everything we know is a meme (e.g. Brodie 1996; Gabora 1997). Brodie includes operant conditioning, and indeed all conditioning, as memetic. Gabora goes even further and counts as a meme 'anything that can be the subject of an instant of experience'. This is extremely confusing. It takes away any power of the idea of the meme as a replicator and adds nothing to the already difficult problem of how to deal with consciousness. If we are going to make progress we need to stick to our clear and simple definition.

What about emotions? Emotions are an inextricable part of human life and even play a key role in rational thought and decision making. The neurologist Antonio Damasio (1994) has worked with many patients who have brain damage, often in the frontal lobe, that causes them to lose their normal emotional responses and become emotionally flat. Far from turning into super-rational decision-makers, able to plan their lives without all the irritating distraction of unwanted emotions, they become almost paralysed with indecision. Whether to choose pickle and pumpkin crisps, or cheese and onion, can become a nerve-racking dilemma to be resolved only by long and careful thought, and a normal life becomes impossible. Most of us would just think 'well, I feel like cheese and onion today' not realising that the emotions have done the complex work of juggling the possible consequences, weighing up the results of past experiences, throwing these in with species-specific preferences and coming up with some rough and ready bodily reaction that allows that tiny verbal part of our brain to say 'I think I'll have the cheese and onion please – if you don't want it'. *Star Trek*'s Mr Data is simply implausible. If he truly had no feelings he would not be able to decide whether to get up in the morning, when to speak to Captain Picard, or whether to drink tea or coffee.

Emotions and thought are intimately linked in other ways too. There are rather few hormones, such as adrenaline and noradrenaline, that control emotional states, but we can experience a vast array of different emotions according to how we interpret and label our physiological responses. In this way you could say that memes come to be involved in our emotions, but are emotions memes? The answer is – only if they can be transmitted to someone else by imitation.

It is almost a truism to say 'you can't possibly know how I'm feeling'. Emotions are private and notoriously difficult to communicate. We write poems, give roses, and paint pictures to try, in some small way, to communicate them. We might, of course, pick up an emotion from someone else, and this certainly looks like imitation, as when tears of sadness spring up in response to seeing another's grief. This contagious spread of behaviour looks like imitation because one person does something and then another person does the same thing. But strictly speaking it is not. To understand why we need to define imitation.

Imitation, contagion, and social learning

The psychologist, Edward Lee Thorndike (1898), was possibly the first to provide a clear definition of imitation as 'learning to do an act from seeing it done'. Thorndike's definition (although confined to visual information) captures the essential idea that in imitation a new behaviour is learned by copying it from someone else. A hundred years later we can see the importance of this point in distinguishing between 'contagion', 'social learning' and 'true imitation'.

The term 'contagion' is used in many different ways. We may think of ideas as contagious, and compare the spread of memes with the spread of infectious or contagious diseases (Lynch 1996). Also the term 'social contagion' is used to describe the spread of behaviours, such as crazes or even suicide, through society (Levy and Nail 1993; Marsden 1998). However, this is not the kind of contagion I want to contrast with imitation. Rather, I mean what has variously been called instinctive imitation, behavioural contagion, social facilitation, co-action, or (simply) contagion (Whiten and Ham 1992). Unfortunately, social psychologists often confuse imitation and contagion, or treat them as the same thing (Grosser *et al.* 1951; Levy and Nail 1993). However, comparative psychologists (those comparing animal and human behaviour) have recently made clear a useful distinction.

Yawning, coughing and laughter are all extremely contagious in humans. Indeed, it can be difficult not to laugh if everyone around you is already laughing. This kind of contagion is thought to rely on specific stimulus feature detectors which detect laughing or yawning in someone else and then trigger the same innate behaviour as the response. In other animals, alarm calls and other vocalisations can be contagious, but contagious laughter appears to be limited to humans (Provine 1996). Other examples include the spread of moods and emotions through crowds, and the way people will stop to look at something if they see other people staring.

This kind of contagion is not true imitation. We can see why by considering Thorndike's simple definition. Yawning, coughing, laughing, and looking are innate behaviours. When we start laughing because everyone else is laughing we have not learned 'how to do an act'. We already know how to laugh, and the kind of laugh we make is not modelled on the laugh we hear. So this kind of contagion is not imitation and should not be counted as memetic.

Then there is social learning (as opposed to individual learning), which is learning that is influenced by observing, or interacting with, another animal or person. Imitation is one form of social learning but there are others that are not truly imitative. Animal researchers have recently made considerable progress in distinguishing between these kinds of learning, and finding out which animals are capable of true imitation (Heyes and Galef 1996). The results have been surprising.

In 1921, in the south of England, tits (small garden birds) were seen prising open the wax-board tops of milk bottles left on the doorstep. Subsequently, the habit became widespread across England and some parts of Scotland and Wales, with other species of bird joining in, and foil tops being pecked as well. That the tits learned from each other was suggested by the way the trick spread gradually from village to village, and across different areas, although it was obviously independently reinvented many times (Fisher and Hinde 1949). With the advent of supermarkets and cardboard cartons, the bottle left by the milkman is becoming rare, but even today you will occasionally find your silver top pierced.

The spread of milk bottle pecking was a simple cultural phenomenon but purists would argue that it was based not on imitation, but on a simpler kind of social learning (Sherry and Galef 1984). Imagine that one bird learned, by trial and error, that there was cream to be had by pecking at the bottles. Then another bird chanced by and saw the pecking and the obviously pecked top. Pecking is a natural action for tits and now that the attention of the second bird had been drawn to the bottle it was more likely to land on it and peck too. Reinforcement in the form of nice tasty cream would lead this bird to repeat the action and possibly be seen by other birds, and so on. The fact that the birds used lots of different methods for opening the bottles also suggests they did not learn by direct imitation.

This kind of social learning is sometimes called 'stimulus enhancement' – the stimulus, in this case the bottle top, has become more readily noticed. Similarly, 'local enhancement' is when attention is directed towards a specific place. Animals also learn from each other which objects or places to fear or ignore. For example, young rhesus monkeys learn to avoid snakes after seeing their parents react fearfully to a snake, and octopuses will attack something they have seen others attacking. Birds and rabbits learn not to fear trains by following others who are not afraid, and therefore become used to the frightening noise. Oystercatchers open mussel shells by either stabbing or hammering techniques according to their tradition, and birds learn to choose migration routes and nesting sites from other birds (Bonner 1980, gives many interesting examples). But none of these processes is true imitation because no new behaviours are passed on from one animal to another (for reviews of social learning and imitation see Heyes and Galef 1996; Whiten and Ham 1992; Zentall and Galef 1988).

Other famous examples that look like true cultural learning based on imitation include the troop of Japanese monkeys that learned to wash sweet-potatoes, and chimpanzees that learned how to fish for termites by poking sticks into the mounds. However, further study of the spread of these skills, and of the animals' learning abilities, suggests that both these traditions depend on individual learning and the kinds of social learning described above, not on true imitation (Galef 1992). So if you want to be really precise about it you have to say that bottle top pecking, termite fishing and sweet-potato washing are not true memes – though they do come close.

What about your neighbourhood blackbird who sings like your alarm clock or imitates a car alarm? True imitation does occur in birds, although their powers of imitation are limited to sounds, and to rather specific kinds of sound at that (with the possible exception of parrots who may be able to imitate simple gestures). For this reason birdsong has long been treated as a special case (Bonner 1980; Delius 1989; Thorndike 1898; Whiten and Ham 1992). Many songbirds have long traditions. The young learn what to sing by imitating their parents or neighbours. In chaffinches, for example, the nestling may hear its father sing long before it is capable of singing itself. A few months later it begins to make a wide variety of sounds, gradually narrowing down to the song it heard as a chick. Experiments show that there is a critical period for learning and that the bird has to hear its own song and match it to the remembered song it is imitating. Hand-raised birds can learn songs from tape recorders and adopted birds sing songs more like their adopted, not biological, parents. Some species learn many songs from neighbours and a few, like parrots and mynahs, can imitate human speech. So we can count birdsong as a meme. Indeed, the cultural evolution of chaffinch song has actually been studied in terms of the mutation, flow and drift of song memes (Lynch *et al.* 1989), and studies of singing honeyeaters have shown that their song meme pool is more diverse on mainland Australia than on a nearby island (Baker 1996). Birdsong is therefore unlike the examples of social learning we were considering before.

The difference can be explained like this. Imitation is learning something about the form of behaviour through observing others, while social learning is learning about the environment through observing others (Heyes 1993). The tits already knew how to peck; they only learned *what* to peck. The monkeys already knew how to be frightened, they only learned *what* to fear.

After nearly a century of research there is very little evidence of true imitation in non-human animals. Birdsong is obviously an exception, and we may be simply ignorant of the underwater world of dolphin imitation. Chimpanzees and gorillas that have been brought up in human families occasionally imitate in ways that their wild counterparts do not (Tomasello *et al.* 1993). However, when apes and human children are given the same problems, only the children readily use imitation to solve them (Call and Tomasello 1995). It seems we are wrong to use the verb 'to ape' to mean imitate, for apes rarely ape.

By contrast, humans are 'the consummate imitative generalist' (Meltzoff 1988, p. 59). Human infants are able to imitate a wide range of vocal sounds, body postures, actions on objects, and even completely arbitrary actions like bending down to touch your head on a plastic panel. By 14 months of age they can even delay imitation for a week or more (Meltzoff 1988), and they seem to know when they are being imitated by adults (Meltzoff 1996). Unlike any other animals, we readily imitate almost everything and anything, and seem to take pleasure in doing so.

If we define memes as transmitted by imitation then we must conclude that only humans are capable of extensive memetic transmission. Some other theorists have included all forms of social learning in their definitions of cultural evolution (e.g. Boyd and Richerson 1985; Delius 1989). Their mathematical models may usefully apply to all; however, I suggest that it will be better for memetics to stick to the original definition of memes. The reason is that the other forms of social learning do not support a replication system with true heredity, because the behaviour is not really copied.

We can think of it this way. In social learning, one animal may invent a new behaviour during individual learning and then somehow lead a second animal into such a situation that it is likely to learn the same new behaviour – or perhaps the first can behave in such a way as to change the contingencies of learning for the second animal so that it learns the same (or a similar) new behaviour. The result looks like copying but it is not, because the behaviour must be created anew by the second learner. The social situation, and the behaviour of the other animal plays a role, but the details of the first behaviour are not transmitted and therefore cannot be built upon and refined by further selective copying. In this sense, then, there is no true heredity. This means there is no new replicator, no true evolution, and therefore the process should not be considered as memetic.

By contrast, the skill of generalised imitation means that humans can invent new behaviours of almost unlimited kinds and copy them on to each other. If we define memes as transmitted by imitation then whatever is passed on by this copying process is a meme. Memes fulfil the role of replicator because they exhibit all three of the necessary conditions, that is, heredity (the form and details of the behaviour are copied), variation (they are copied with errors, embellishments or other variations), and selection (only some behaviours are successfully copied). This is a true evolutionary process.

....

We have now established that imitation is rare and special, but just what is entailed in an act of imitation? There is considerable research on imitation in infants and children (Meltzoff and Moore 1977; Whiten *et al.* 1996; Yando *et al.* 1978), and some on sport, on social conformity, and on questions like whether violent television causes copycat violence (Bandura and Walters 1963) and whether suicide, vehicle accidents, and even murder can spread by imitation (Marsden 1998*b*; Phillips 1980). However, there is little on the mechanisms underlying imitation, and so I will have to speculate a little.

We might liken the process to 'reverse engineering', a common way of stealing ideas in modern industry. If an unscrupulous manufacturer wants to make a cheap version of the latest high-tech compact disk player then specially trained engineers tear the real thing apart bit by bit, trying to work out what all the parts do and how they can be made. With luck they can then build their own version to perform the same way – without paying royalties. But it is not easy.

Now imagine you are going to copy a simple action. Suppose I put my hands to my mouth in a trumpet-like shape, point them upwards and hum 'de-tum-de-tum'. I would bet that, unless you were physically unable, you would have little trouble in copying me – and that people watching would agree on whether you managed a good performance or not. What is so difficult about that?

Everything. First, you (or rather some unconscious brain mechanisms) have to decide which aspects of the action are to be copied – does the angle of your leg matter? or the position of your feet? Is it more important that your hands look something like a trumpet or that their exact position is as close as possible to my version of a trumpet? Must your humming be in the same key, or only follow the same melody? I am sure you can make up your own questions. Having decided on the important aspects to be copied, a very difficult set of transformations has to be effected. You watched me, let's say, from the side. Nothing you saw of my actions will correspond to the way the actions will look from your perspective when you carry them out yourself. You will see only your hands from the near end of the 'trumpet'. Somehow, your brain has to create a transformation of the action I did that will enable it to instruct your muscles to do whatever they have to do to get your action to look like mine to someone else. Now it begins to sound complicated.

It sounds complicated because it is. Imitation necessarily involves: (a) decisions about what to imitate, or what counts as 'the same' or 'similar', (b) complex transformations from one point of view to another, and (c) the production of matching bodily actions.

Once you realise how difficult this natural-seeming kind of act is, it is tempting to think we cannot possibly do it – although obviously we *do* do it. Or that a science of memetics cannot be based on something so peculiar. I am reassured by simply reminding myself that human life really is like this. We *do* copy each other all the time and we underestimate what is involved because imitation comes so easily to us. When we copy each other, something, however intangible, is passed on. That something is the meme. And taking a meme's eye view is the foundation of memetics.

CHAPTER 5

Three problems with memes

Is Beethoven's *Fifth Symphony* a meme, or only the first four notes?

This raises a real question for memetics and one that is worth exploring – but I do not think it is a problem. There are several such objections to memetics that are frequently raised and worth trying to resolve. I am going to consider three and will argue that all are either soluble or irrelevant.

We cannot specify the unit of a meme

Whether by coincidence or by memetic transmission, Beethoven is the favourite example for illustrating this problem. Brodie (1996) uses Beethoven's *Fifth Symphony*, Dawkins (1976) uses the *Ninth*, and Dennett (1995) uses both the *Fifth* and the *Seventh*. Dennett adds that the first four notes of Beethoven's *Fifth* are a tremendously successful meme, replicating all by themselves in contexts in which Beethoven's works are quite unknown. So are they the meme, or the whole symphony?

If we cannot answer this question we cannot identify the unit of the meme, and some people clearly think this is a problem for memetics. For example, many years ago Jacob Bronowski wondered why we do not have a better understanding of social change and blamed our not being able to pin down the relevant units (Hull 1982). I have heard people dismiss the whole idea of memetics on the grounds that 'you can't even say what the unit of a meme is'. Well that is true, I cannot. And I do not think it is necessary. A replicator does not have to come neatly parcelled up in ready-labelled units. Since genes are our most familiar example we should look at the same issue for them.

Defining a gene is not easy and in fact the term is used quite differently by breeders, geneticists and molecular biologists because they are interested in different things. At the molecular level, genes consist of sequences of nucleotides along a molecule of DNA. Names are given to different lengths of DNA, such as a codon, which is a sequence of three nucleotides, or a cistron, which is a sufficiently long sequence of nucleotides to provide the instructions for building one protein – with a start symbol and a stop symbol. Neither of these is necessarily passed on intact in sexual reproduction and neither corresponds with what we think of as the gene 'for' something. DNA provides instructions for protein synthesis and it is a long way from there to having blue or brown eyes, finding men more sexy than women, or having a flair for music. Yet it is these effects of genes that natural selection gets to work on. So what is the unit of the gene?

Perhaps there is no final answer. One useful suggestion is that a gene is hereditary information that lasts long enough to be subject to the relevant selection pressures. A sequence of DNA that is too short is meaningless – it lasts almost indefinitely, being passed on identically from generation to generation but taking part in countless different kinds of protein synthesis and countless different phenotypic effects. A sequence that is too long does not survive through enough generations to be selected for or against. So some intermediate length has to be chosen, and even this varies with the strength of the selection pressure (see Dawkins 1976; Williams 1966).

This intrinsic uncertainty about just what to count as a gene has not impeded progress in genetics and biology. It has not made people say, 'We cannot decide what the unit of the gene is so let's abandon genetics, biology and evolution.' These sciences all work by using whatever unit they find most helpful for what they are doing at the time.

The same logic applies in memetics. Dennett (1995) defines the units of memes as 'the smallest elements that replicate themselves with reliability and fecundity' (p. 344). A blob of pink paint is too small a unit for memetic selection pressures to apply – to be enjoyed or disliked, photographed or thrown away. A whole gallery of paintings is too large. The single painting is the natural unit for most of us and that is why we remember Van Gogh's *Sunflowers* or buy postcards of Edvard Munch's *The Scream.* Styles of painting, such as impressionism or cubism, can also be copied and therefore count as memes, but can hardly be divided up into units. A single word is too short to copyright and an entire library too long, but we can and do copyright anything from a clever advertising jingle to a 100 000-word book. Any of these can count as memes – there is no right answer to the question – 'What *really* is the unit of the meme'.

I might have argued that four notes is too short to be a meme but everyone's favourite example shows I am wrong. If a musical genius picks on just the right four notes, starts a wonderful symphony with them and has the luck to have his work survive into an age of mass communication – then his four notes can be heard and remembered by literally billions of people. I am sorry if you are one of them and now are not able to get these four notes out of your mind.

This problem – why can't I get that tune out of my mind – provides a good example of memetics at work, and I shall use it to show that the size of the unit makes no difference.

Why do tunes sometimes just go round and round in my head and will not go away? Why do we have the sort of brains that do that? What possible use is it for me to spend all day singing 'Coke Refreshes you Best', or the theme tune from *Neighbours*? The answer from memetics is that it is no use at all to me – but it is of use to the memes.

Memes are replicators and if they can get themselves copied they will. The imitating machinery of the brain is an excellent environment for copying tunes. So if a tune is memorable enough to get lodged in your brain and then passed on again then it will – and if it is really memorable, or singable, or playable, it will get into a lot of brains. If it turns out to be just what some TV producer needs to start her latest soap opera then it will get into even more brains, and every time you start humming it there is a chance that someone else will hear you and you will set them off. Meanwhile, plenty of other tunes are never heard again. The consequence of all this is that the successful ones increase in the meme pool at the expense of the others. We all get infected with them and they are stored in our memories, ever ready to be activated and passed on to anyone who has not got them yet. All this singing is not for our benefit, nor for our genes' benefit. Being haunted by horrible tunes is just an inevitable consequence of having brains that can imitate tunes.

Note that this argument works regardless of all the specific reasons why one piece of music may be singable or likeable and another not. Those reasons might include, for example, innate preferences for certain sounds, the pleasure to be found in predictability and unpredictability of sounds, or overall complexity. Gatherer (1997) has explored the development of jazz in terms of the adaptiveness of its component parts, looking at complexity, memorability and the effect of the available technology at different times. Simple melodies are easy to remember but may not be interesting enough for people to pass on. Complex improvised music can evolve but may only survive in a community of trained musicians and listeners, while even more complex music may be simply too difficult to remember and so fail to be replicated, even if it can be enjoyed. Memetics in the future may discover what makes for successful replication in music. It may find out how different kinds of music fill different niches, such as specialised replication among minority groups, or short-lived mass popularity. But note that none of this matters for the simple argument I am making here. That is, that any catchy tune that gets you to rehearse it in your head will get passed on, and so we will all come across such tunes and be in danger of 'catching' them.

Memetics thus provides a simple and obvious explanation for those irritating tunes that go round and round in our heads – as it did for why we cannot stop thinking in general. The tunes are like weeds and just tend to grow. Does it matter what we count as the unit of a meme in either of these cases? I say no. The competition to grab any spare brain power will go on regardless of the way we might decide to divide the competing instructions. The meme is 'whatever it is that is passed on by imitation'. If your irritating humming at work passes on both verses of Blake's Jerusalem to the rest of the office then the whole inspirational song is the meme. If you infected them only with 'Da, da, da, dum', then those good old four notes are the meme.

We do not know the mechanism for copying and storing memes

No we do not. The fact that we now know so much about how DNA works can easily lead us to imagine that we need that level of understanding for memetics – right away. I do not think this is so. Don't forget how far evolutionary theory got before DNA was even heard of. Darwin's *Origin of Species* was published in 1859. It was not until the 1930s that genetics and natural selection were brought together (Fisher 1930); not until the 1940s that other areas of science were brought into what is sometimes known as the modern synthesis, leading to neo-Darwinian theory; and not until the 1950s that the structure of DNA was finally discovered (Watson 1968). In the first century of Darwinism an enormous amount was achieved in the understanding of evolution without anyone having any idea about chemical replication, the control of protein synthesis or what on earth DNA was doing.

The memetics we build at the very end of the twentieth century will doubtless appear inept in another century's time, but that is no reason not to begin. We may get a long way with the general principles of memetic selection without understanding the brain mechanisms it relies on. We can also make some educated guesses about those mechanisms based on what little we do know.

First, we may assume that, at least at some phase in their replication, memes have to be physically stored in brains. As far as storage is concerned neuroscience is making great strides in working out the biological basis of memory. Artificial neural networks have demonstrated that many of the features of human memory can be simulated in computers. Work on synaptic transmission, long-term potentiation and neurotransmitters is finding out whether real brains do anything similar. If they do, we may guess that human memory probably works something like this (for example, see Churchland and Sejnowski 1992).

Neural networks in the brain consist of large agglomerations of individual cells with a layer of cells taking input (e.g. from the eyes, or from another network), another layer providing output (e.g. to the muscles, the voice, or another network), and many layers in between. Each neuron has connections to many others and the strength of these connections varies according to their history. At any given state of the network a certain kind of input will produce a certain kind of output but this relationship is not fixed. The network can be trained, for example by consistently pairing certain sorts of input, and this experience changes its responses to new inputs. In other words, it can remember.

This kind of memory is nothing like the memory of a digital computer, with its fixed locations, nor like a tape recorder with its more or less faithful duplication of everything fed into it. In a brain, every input builds on what went before. In a lifetime of complex experiences we do not store each one away in a black box to be retrieved later when we need it – rather, every experience comes into a complex brain and has a greater or lesser effect on what it finds there. Some things have virtually no effect and are completely unmemorable (we could not function otherwise). Some have enough of an effect to stay briefly in short-term memory but are then lost, while some lead to dramatic changes so that precise events can easily be reconstructed, whole poems recited or that special face never forgotten.

Effective memes will be those that cause high fidelity, long-lasting memory. Memes may be successful at spreading largely because they are memorable rather than because they are important or useful. Wrong theories in science may spread simply because they are comprehensible and fit easily with existing theories, and bad books may sell more copies because you can remember the title when you get to the bookshop – though, of course, we do have strategies for overcoming these biases. An important task of memetics will be to integrate the psychology of memory with an understanding of memetic selection.

Some people argue that memes are not digital (Maynard Smith 1996) and that only digital systems can support evolution. Certainly genes are digital and certainly digital storage is far preferable to analogue. We all know that digital video- and audio-recordings look and sound better than their analogue predecessors; a digital system allows information to be stored and transmitted with far less loss of information even over noisy channels. However, there is no law that says that evolution has to be digitally based – the issue is really one of the quality of replication.

What, then, makes for a good quality replicator? Dawkins (1976) sums it up in three words – fidelity, fecundity, and longevity. This means that a replicator has to be copied accurately, many copies must be made, and the copies must last a long time – although there may be trade-offs between the three. Genes do well on all three counts, and being digital gives them high fidelity copying. So what about brains?

Our memory is obviously good enough for us to learn several languages, recognise thousands of photographs from one showing, and recall the major events of our lives over periods of decades. Is this good enough to support memetic evolution? I think this is an empirical question that could be put to the test. In the future, memeticists might be able to develop mathematical models to determine just how high the fidelity of memory must be to support memetic evolution, and compare that with the known performance of human memory. My guess is that our memories will be found to be quite good enough, whether they ultimately turn out to be digital or not.

Second, memes depend on being transmitted from one person to another and, by definition, this is done by imitation. We have already seen how poorly understood imitation is but we may at least make a simple prediction. Actions that are easy to imitate will make for successful memes and ones that are difficult to imitate will not.

Apart from that, the effective transmission of memes depends critically on human preferences, attention, emotions and desires – in other words, the stuff of evolutionary psychology. For genetic reasons we are driven by the desire for sex, for sex of different kinds, for food, for better food, for avoiding danger and for excitement and power. Evolutionary psychology already provides us with lots of information that explains why some memes are picked up again and again while others make no impact. We need to use that information and build on it.

To conclude – it is true that we do not understand in detail how memes are stored and transmitted. But we have plenty of clues and we certainly know enough to get started.

Memetic evolution is 'Lamarckian'

Biological evolution is not Lamarckian and cultural evolution is – or so I have heard. This apparent difference has been frequently noted, and many treat it as a problem (Boyd and Richerson 1985; Dennett 1991; Gould 1979, 1991; Hull 1982; Wispé and Thompson 1976). In a recent discussion of artificial life, the British biologist John Maynard Smith asked what features are necessary for any evolving system – natural or artificial – and suggested 'digital coding and non-Lamarckian inheritance' (Maynard Smith 1996, p. 177). So is memetic evolution really Lamarckian? And what would be the significance for memetics if it were?

First, the term 'Lamarckian' has come to refer to just one aspect of the evolutionary theory of Jean-Baptiste de Lamarck. Lamarck believed in all sorts of things that have now been rejected, including the inevitability of progress in evolution and the importance of organisms striving towards their own improvement. However, what is now referred to as 'Lamarckism' is the principle of the inheritance of acquired characteristics. That is, if you learn something or undergo some change during your lifetime, you can pass it on to your offspring.

Lamarckism (in this sense) is not true of biological evolution, at least in sexually reproducing species. The way inheritance works (which was not understood in Darwin's or Lamarck's time) makes it impossible. This is sometimes known as 'Weismann's barrier' after August Weismann who, at the end of the nineteenth century, pointed out what he called the 'continuity of the germ-plasm'. In more modern terms we can see it like this – using the example of sexual reproduction and human beings.

The genes are coded in DNA and stored in pairs of chromosomes in every cell of your body. At any location on a chromosome different people may have different alleles (versions) of the same gene and the total make-up of genes in each individual is known as their genotype. Correspondingly, the various characteristics of the final person is known as the phenotype. The genes are not a blue print or a map of the future phenotype; they are instructions for building proteins. These instructions control the development of the embryo as it grows, and of the adult as it develops in its own unique environment. The result is a phenotype which is highly dependent on the genotype it started from but is in no sense a copy of that genotype or completely determined by it.

Now imagine that you acquire some new characteristics by, say learning a language, practising playing the piano, or building up your thigh muscles – that is, your phenotype changes. There is no way that this change in your body can affect the genes that you pass on – although it can affect *whether* you pass some on or not. The genes your children will inherit are derived directly from the genes you inherited and it is this continuous line that is known as the 'germ line'. Conceivably, if genes worked as some kind of stored blueprint or map then changes in the phenotype could be fed back to change the map but this is not the case. Conceivably, the process of meiosis, in which cells divide to make ova and sperms, could be affected by changes in the phenotype, but this does not happen and in any case, the ova a woman carries in her ovaries are already there at her birth. We must imagine the germ line going on continuously, with the genes being shuffled and recombined in each generation. These genes instruct the phenotypes which then set off on their own and are either successful or not, but the phenotypes do not instruct the genes.

Even though Lamarckian inheritance cannot happen in such a system, there have been many experiments looking for it. Weismann himself cut off the tails of mice for many generations with no obvious effect on the length of their offspring's tails. However, this is not strictly a test of the theory because Lamarck argued that organisms had to strive towards improvements, as when giraffes stretch their necks, or birds practise flying, and Weismann's mice presumably did not strive to have their tails cut off. In Russia, the official science of Lysenko was based on Lamarckism but produced no progress in biology and was disastrous for Soviet agriculture because their plant breeding programmes failed.

Lamarck's idea is still popular and appears in many guises, including memories of past lives being attributed to 'genetic memory', and psychic powers being explained by 'spiritual evolution'. Perhaps it is popular because it implies that there is some point in all our hard work or some benefit to our children if we struggle to improve ourselves. But from the purely genetic point of view there is no such benefit. Popular it may be, but it is simply not true.

At least, Lamarckism is not true for sexual species. For other kinds of organism the idea is simply inapplicable. The most common creatures on this planet are unicellular organisms, such as bacteria, which reproduce by cell division. For these ubiquitous creatures there is no clear distinction between genotype and phenotype, genetic information is exchanged in various ways, and there is no clear germ line. So the whole idea of Lamarckian inheritance is irrelevant.

What then of cultural evolution? The answer depends critically on how you draw the analogy between genes and memes and, as I have stressed before, we must be very careful whenever we use this analogy.

One way to draw the analogy is to stick to the notion of the human genotype, phenotype and generation. In this case, acquired characteristics are certainly passed on, as when religions are transmitted from parents to children generation after generation. But memes do not stick to biological generations and can jump about all over the place. If I invent a brilliant new recipe for pumpkin soup, I can pass it on to you and you can pass it on to your granny and she can pass it on to her best friend. Also, this is not inheritance in the biological sense and the genes are not affected. So it is not Lamarckian.

A more interesting way to use the analogy is to forget about phenotypes and biological generations and look at memes and memetic generations. In the case of the soup there were three generations between me and your granny's best friend. In each generation, the recipe went from brain to behaviour in the kitchen and on to the next brain (that is if you watched me make the soup). Was there inheritance of acquired characteristics? Let us say that the meme in my brain is the equivalent of the genotype and my behaviour in the kitchen is the equivalent of the phenotype. Then, yes, inheritance is Lamarckian because if I put in too much salt on this occasion, or you forget one of my special herbs, or fail to copy the way I shred the garlic, then you will pass on this new version when your granny watches you, and so the new phenotype will have acquired the characteristics accordingly.

But what if you did not watch me make the soup? What if I sent you the recipe in the post and you passed it on to your granny and she made a photocopy for her friend? Now the situation is quite different. We may draw the analogy with biology this way. The written recipe is like the genotype, it contains the instructions for making the soup. The soup is like the phenotype. The delicious taste of the soup is the reason why the recipe is copied – your granny only asked for a copy of the recipe because she liked the soup. In this case, if she fails to follow the recipe correctly her alterations may affect the chance of someone else wanting the recipe, but the alterations will not be passed on because they are only in the soup itself (phenotype) not in the written recipe (genotype). In this case the process is perfectly analogous to the biological situation and is *not* Lamarckian.

I shall call these different modes of transmission 'copy-the-product' and 'copy-the-instructions'. Music provides a slightly different example. Let us suppose that my daughter plays a beautiful piece of music for her friends and one of them wants to learn to play it too. Emily could either play it many times until her friend can copy it accurately (copy-the-product), or simply hand her the written music in a book (copy-the-instructions). In the first case, any changes Emily makes will be passed on, and if there follows a series of pianists copying each other, the composition may gradually change, incorporating the errors or embellishments of each player. In the second case, the individual playing styles of each pianist will not have any effect because copies of the (unembellished) written music are passed on. In the first case the process appears Lamarckian but in the second case it does not.

In the biological world, sexual species work by copying-the-instructions. The genes are the instructions that are copied, the phenotype is the result and is not copied. In the world of memes, in which both processes are used, you could argue for calling 'copy-the-instructions' Darwinian, and 'copy-the-product' Lamarckian, but I suggest this would only lead to more confusion. I deliberately described the soup and the music in ways that allowed the two modes of replication to be easily separated, but in the real world they may be inextricably mixed. From me to your granny's friend the instructions on making the soup might go from brain to piece of paper, to behaviour, to another brain, to a computer disk and another piece of paper and to another brain – with lots of different flavoured soups being made along the way. Which is the genotype and which the phenotype in each case? Are we to count memes as only the instructions in the brains or the ones on paper too? Are the behaviours memes or meme-phenotypes? If the behaviour is the phenotype, what then is the soup? There are lots of possibilities in memetic evolution because memes are not confined by the rigid structure of DNA. The ways they spread are legion. But we can only decide whether memetic evolution is *really* Lamarckian if we can answer these questions. We seem to be at an impasse.

Fortunately, we need not worry. All this trouble is caused by expecting there to be a close analogy between memes and genes when there need not be. We must remember Campbell's Rule and the basic principle of memetics – that genes and memes are both replicators but otherwise they are different. We need not, and must not, expect all the concepts from biological evolution to transfer neatly across to memetic evolution. If we do we will hit trouble as we have done here.

My conclusion apropos Lamarck is that the question 'Is cultural evolution Lamarckian' is best not asked. The question only makes sense if you draw certain kinds of strict analogy between genes and memes but such analogies are not justified. We are best to confine the term 'Lamarckian' to discussion of biological evolution in sexually reproducing species. When we come to other kinds of evolution the distinction between mechanisms that 'copy-the-instructions' and those that 'copy-the-product' will prove more helpful.

Terminology

So what do we call that soup? The value of asking the question about Lamarck is that it does make us face up to some really tricky questions about terminology. Some previous authors have, understandably, evaded these questions, while others have launched in and made distinctions that might turn out not to be justified. In fact, the terminology of memetics is in a mess and needs sorting out. I am going to consider the use of three terms: meme, meme-phenotype (sometimes called phemotype), and meme vehicle.

First, what are we to count as a meme? In the case of the soup, is it the stored instructions in my brain, the soup itself, my behaviour in the kitchen, the words on the piece of paper, or all or none of these things? We might have doubts about the soup because, however delicious it is, you could not easily work out how it was made from tasting it – though perhaps an expert chef might be able to do it, just as a musician might be able to reconstruct a piece of music from hearing it. So do we need a different scheme for copyable meme products from uncopyable ones? I am deliberately making life difficult for myself because no consensus has yet emerged and if memetics is going to make progress we will have to agree on fundamentals like this. Let us see whether definitions exist that can help us sort it all out.

Dawkins (1976) initially did not commit himself at all and used the term 'meme' to apply to the behaviour, the physical structure in a brain, and memetic information stored in other ways. His original examples, remember, were tunes, ideas, catchphrases, clothes fashions, and ways of making pots or arches. Later he decided that 'A meme should be regarded as a unit of information residing in a brain (Cloak's i-culture)' (Dawkins 1982, p. 109). This implies that the information in the clothes or the arches does not count as a meme. But later still he says that memes 'can propagate themselves from brain to brain, from brain to book, from book to brain, from brain to computer, from computer to computer' (Dawkins 1986, p. 158). Presumably, they still count as memes in all these forms of storage – not just when they are in a brain.

Dennett (1991, 1995) treats memes as the ideas that are passed on; whether they are in a brain or a book or some other physical structure, they are information undergoing the evolutionary algorithm. He points out that the structure of a meme may not be the same in any two brains – indeed it almost certainly will not be – but when a person carries out any behaviour there must be some kind of instruction stored in their brain, and when someone else copies and remembers an action they must also create some kind of neural change. Durham (1991) also treats memes as information, again regardless of how it is stored.

In contrast, Delius (1989) describes memes as 'constellations of activated and non-activated synapses within neural memory networks' (p. 45), or 'arrays of modified synapses' (p. 54). Lynch (1991) defines them as memory abstractions and, in his memetic lexicon, Grant (1990) defines memes as information patterns infecting human minds. Presumably, on these latter definitions, memes cannot be carried by books or buildings, and the books and buildings must be given some other role. This has been done, by using further distinctions.

The usual way to make the distinction is, of course, by analogy with genes. A common one is to use the concept of the phenotype. Cloak (1975) was the first to do this and was very clear about it. He defined the i-culture as the instructions in people's heads, and the m-culture as the features of people's behaviour, their technology and social organisation. He explicitly likened his i-culture to the genotype and m-culture to the phenotype. As we have seen, Dawkins initially did not make such a distinction, but in *The Extended Phenotype* he says 'Unfortunately, unlike Cloak . . . I was insufficiently clear about the distinction between the meme itself, as replicator, on the one hand, and its "phenotypic effects" or "meme products" on the other' (Dawkins 1982, p. 109). He then went on to describe the meme as the structure physically realised in the brain.

Dennett (1995) also talks about memes and their phenotypic effects, but in a different way. The meme is internal (though not confined to brains) while the design it shows the world, 'the way it affects things in its environment' (p. 349), is its phenotype. In an almost complete reversal, Benzon (1996) likens pots, knives, and written words (Cloak's m-culture) to the gene, and ideas, desires and emotion (i-culture) to the phenotype. Gabora (1997) likens the genotype to the mental representation of a meme, and the phenotype to its implementation. Delius (1989), having defined memes as being in the brain, refers to behaviour as memes' phenotypic expression, while remaining ambiguous about the role of the clothes fashions he discusses. Grant (1990) defines the 'memotype' as the actual information content of a meme, and distinguishes this from its 'sociotype' or social expression. He explicitly bases his memotype/sociotype distinction on the phenotype/genotype distinction.

Although these ideas have something in common they are not all the same and it is not at all clear, at least to me, which is better. On the whole, I think none of them really works because they have not appreciated the difference between the copying-the-product and copying-the-instructions.

The notion of a phenotype applies easily to one but not to the other and there may be other modes of transmission as well. I will not, therefore, use the concept of the meme-phenotype because I cannot give it a clear and unambiguous meaning.

Another analogy is made with the concept of the vehicle. Dawkins (1982) originally introduced the distinction between replicators and vehicles in the context of genetic selection, in order to make clear that the genes are the selfish replicators, while it is much larger units – typically (though not necessarily) whole organisms – that either live or die. He described organisms as vehicles for the genes, built to carry them around and protect them. Dawkins defines a vehicle as 'any unit, discrete enough to seem worth naming, which houses a collection of replicators and which works as a unit for the preservation and propagation of those replicators' (p. 114).

Using this concept, Dennett treats memes as ideas and the physical objects that carry them around as meme vehicles. So, for example, 'A wagon with spoked wheels carries not only grain or freight from place to place; it carries the brilliant idea of a wagon with spoked wheels from mind to mind' (Dennett 1995, p. 348 and 1991, p. 204). Pictures, books, tools and buildings are all meme vehicles for Dennett, and he explicitly makes the comparison with gene vehicles. Brodie (1996) follows Dennett and uses the term 'vehicle' for physical manifestations of a meme, as do others. However, there are problems with this analogy (Speel 1995). A wagon may indeed carry around the idea of spoked wheels but does it house a collection of replicators? Does it work as a unit for the preservation and propagation of its memes? A book may seem very much like a vehicle in this sense, but my pumpkin soup does not. I am not at all sure where to draw the lines here.

We must avoid the temptation of assuming there must always be a vehicle, and therefore forcing the memes to fit. Dawkins says he coined the 'vehicle' not to praise it but to bury it. There is no necessity for vehicles to form, and in many kinds of evolution they may not. We should not ask, 'What is the vehicle in this situation,' but, 'Is there a vehicle in this situation and, if so, why?' (Dawkins, 1994, p. 617). We might therefore ask whether memes do in fact group together to make 'a unit for the preservation and propagation of those replicators' and if so what would these true meme vehicles look like? Large self-preserving memeplexes such as religions, scientific theories, or political ideologies might fit the analogy better than wagons and recipes but obviously the term 'vehicle' is being used in quite a different sense here. Finally, the term 'vehicle' can be used in the very ordinary sense that people carry both genes and memes around with them and thus act as their 'vehicles'.

I have thought long and hard about these distinctions. I have tried to see which works well and which does not and so adopt one or other version. I have tried to make new ones of my own and despaired of it. In the end, I come back to what I have called the most basic principle of memetics – that genes and memes are both replicators but otherwise they are different. The analogy between genes and memes has led many people astray and will probably continue to do so for a long time to come. There is an analogy there but only because both are replicators. Beyond that the analogy is weak. There need be no exact memetic equivalent of the phenotype or the vehicle, any more than there are equivalents for strictly genetic concepts like alleles, loci, mitosis and meiosis. In biological evolution genes build their phenotypes but copy themselves straight down through the germ line, but in memetic evolution it can be more like a zigzag with memes hopping from brain to paper to computer and back to brain.

The conclusion I have come to from all of this, is to keep things as simple as possible. I shall use the term 'meme' indiscriminately to refer to memetic information in any of its many forms; including ideas, the brain structures that instantiate those ideas, the behaviours these brain structures produce, and their versions in books, recipes, maps and written music. As long as that information can be copied by a process we may broadly call 'imitation', then it counts as a meme. I shall use the term 'vehicle' only in the ordinary sense of carrying something around, and I shall not use terms like 'sociotype' or 'meme-phenotype' at all. If it turns out later that we need more terms and distinctions then I am sure someone will provide them. It will be easier for someone else to add on necessary distinctions at a later date than to demolish any unhelpful ones that I make now.

This has been a long struggle through some (and certainly not all) of the problems of memetics but I think it will stand us in good stead. Using the simple scheme we have arrived at, and bearing in mind the lurking dangers, we can get on with exploring just what a science of memetics can do – like explaining why we humans have such big brains.

CHAPTER 6

The big brain

The human brain is enormous. Why? Nobody knows for sure. Certainly there have been many theories of the origins of the huge human brain, but still none is universally accepted and a mystery remains. Most theorists assume that the big brain must have evolved by natural selection, such as the American neuroscientist and anthropologist Terrence Deacon (1997) who says 'It cannot be doubted that such a robust and persistent trend in brain structure evolution reflects forces of natural selection' (p. 344) – but if so we must be able to identify the selection pressures involved. So what are they? The answer is not obvious, and the explanatory task to be performed is great. It is basically this.

Origins of the human brain

Human brains today are capable of extraordinary feats quite beyond the abilities of any other species on the planet. Not only do we have language but we have invented fridge-freezers, the internal combustion engine and rocket technology; we can (well, some of us can) play chess, tennis and *Mega-Death 6*; we listen to music, dance and sing; and we have created democracy, social security systems and the stock market. What possible evolutionary advantage could these things have? Or more precisely what selective advantage could there have been for a brain capable of such things? We seem to have a brain 'surplus to requirements, surplus to adaptive needs' (Cronin 1991, p. 355).

In Darwin's time this question so vexed Alfred Russel Wallace that, despite having independently discovered the principle of natural selection, he concluded that it could not account for man's higher abilities (Wallace 1891). Primitive hunter-gatherers could not possibly have needed brains such as these, he reasoned, so there must have been some kind of supernatural intervention. Wallace supported the spiritualists who were claiming to be able to communicate with the surviving spirits of the dead, while Darwin fought against them. Wallace believed that the intellectual and spiritual nature of man was so far above that of the animals that we were different in kind from them. Although our bodies were developed by continuous modification of ancestral animals, some different agency was required to explain our consciousness, morality and spiritual nature, 'the higher feelings of pure morality', courageous self-sacrifice, art, mathematics and philosophy.

Appealing to God or spirits to solve mysteries is no help and few, if any, scientists now favour Wallace's 'solution'. Nevertheless, this old argument highlights a real problem; our abilities are quite out of line with those of other living creatures and they do not seem obviously designed for survival.

The gulf is obvious in purely physical measurements (Jerison 1973). The modern human brain has a volume of about 1350 cubic centimetres (i.e. roughly three times as large as the brains of existing apes of comparable body size). A common way of comparing brain sizes is to use the 'encephalisation quotient' which compares a given animal's brain-to-body ratio with the average for a group of animals. For any group of related animals a plot of brain size against body size yields a roughly straight line (on a log-log scale). If we humans are placed on such a line with our closest living relatives we just do not fit. Our encephalisation quotient compared with other primates is 3. Our brains are far too large for our bodies.

Of course, the encephalisation quotient is only a crude measure and hides the different ways in which the body size to brain size ratio can come about. For example, a chihuahua has a very high encephalisation quotient compared with that of a Great Dane, but this is because chihuahuas have been specially bred for small bodies – not for large brains or superior intelligence! So could we have been selected for small bodies rather than large brains? Deacon (1997), who pointed out the 'Chihuahua Fallacy' explains that the higher encephalisation quotient of primates compared with other animals is a result of their having smaller, slower growing bodies. Primates' brains grow at the same rate as other species' but their bodies grow more slowly. However, when you compare humans with other primates the situation is different. Human fetuses start growing the same way as other primates' but then our brains continue to grow for longer. So our brains do seem to have been selected for extra growth. Our high encephalisation has come about first from the slowed body growth of primates and second from the extra brain growth of humans.

When in evolution did this brain growth begin? About five million years ago the evolutionary branch leading to modern humans split off from that leading to the present day African apes (Leakey 1994; Wills 1993). After this, our early hominid ancestors include various species of australopithecines and then of *Homo* – including *Homo habilis*, *Homo erectus*, and most recently *Homo sapiens*.

The australopithecines include the famous skeleton Lucy, an example of *Australopithecus afarensis* found in Ethiopia by Maurice Taieb and Donald Johansen and named after the Beatles' song Lucy in the Sky with Diamonds. Remains of *A. afarensis* range from four million to less than two and a half million years old. Lucy herself is thought to have lived a little over three million years ago, was about three feet tall and rather ape-like in build with a brain of about 400–500 cc – not much larger than a modern chimpanzee's. From fossil footprints and computer simulations of walking based on fossil bones, it is now clear that *A. afarensis* must have walked upright, though probably could not run. So we know that bipedalism came long before hominid brains began to grow significantly in size.

The increase in brain size probably began about two and a half million years ago, at about the same time (archaeologically speaking) as the beginnings of stone tools and the transition from *Australopithecus* to *Homo*. At this time, global cooling was transforming much of Africa's lush forest into woodland and then into grassy savannah. Adaptation to this new environment is thought to account for some of the changes leading to *Homo*. The first species of *Homo* was *Homo habilis*, named the 'handyman' because of the primitive stone tools they made. Australopithecines may have used sticks or stones that they found as tools, as other apes do today, but *H. habilis* was the first to chip stones into specific shapes to use as knives, choppers or scrapers. Their brains were significantly larger than australopithecine brains at about 600–750 cc.

About 1.8 million years ago *Homo erectus* begin to appear in the fossil record in Kenya. *Homo erectus* was taller and had yet bigger brains, about 800–900 cc. They were the first hominids to travel out of Africa, the first to harness and use fire, and they survived in some parts of the world until as recently as 100 000 years ago. More recently still, the fossil record becomes much richer but there are many arguments about the origins of fully modern humans. So-called archaic *Homo sapiens* are widely distributed and have brains around 1100 cc, with somewhat protruding faces and heavy brow ridges, but there are two main types. One type, which seems to have led to modern *H. sapiens*, appeared in Africa about 120 000 years ago. The other lived at the same time and finally died out only about 35 000 years ago – they were the Neanderthals, *Homo sapiens neanderthalensis*. They had large brow ridges and protruding faces. Their brains were possibly even larger than ours and there is increasing evidence of their use of fire, their culture, and the possibility that they too had language. There is still much argument about which hominid line produced modern humans, and what happened to the Neanderthals. However, sequencing of mitochondrial DNA suggests that they were not our ancestors (Krings *et al.* 1997). So did we kill them off, as we have killed off so many other species, or did they become extinct for some other reason?

A rather odd fact is that for most of the past 5 million years there have always been several species of hominid living at the same time, as there are several species of other primates now. Today there is only one kind of human with rather minor differences around the world. What happened to all the rest?

These are fascinating issues but we must return to our main argument. Most relevant is that brain size increased dramatically during the relatively short period of 2.5 million years that separated the last australopithecines from fully modern humans. By about 100 000 years ago all living hominids would probably class as *H. sapiens* and had brains about as large as ours.

This massive increase must have been very expensive in energy terms. First, the brain is expensive to run. It is often said that the brain consumes 20 per cent of the body's energy but consists of only 2 per cent of the body weight. This figure is slightly misleading because it refers to a body at rest. When large muscles are lugging you and your suitcase as fast as you can go across the platform as the train whistle blows, the brain's energy use is small by comparison. Nevertheless, your muscles often rest, but the brain does not, even in sleep. It uses roughly the energy consumed by a light bulb, all the time.

The brain consists primarily of neurons that conduct impulses along their axons. These impulses consist of a wave of depolarisation which sweeps along the axon as charged ions flood across the axon's membrane. Much of the energy the brain uses is consumed in maintaining the chemical differences across these membranes so that the neurons are continuously ready to fire. Also, many neurons keep firing at a low frequency all the time so that incoming signals can pass on information by either increasing or decreasing the resting frequency. The body's energy budget must have to find a large surplus to keep all this going. A smaller brain would certainly save a lot of energy, and evolution does not waste energy for no reason. As Steven Pinker (1994, p. 363) said 'Why would evolution ever have selected for sheer bigness of brain, that bulbous, metabolically greedy organ? . . . Any selection on brain size itself would surely have favored the pinhead'.

Second, the brain is expensive to build. The neurons are surrounded by a fatty sheath of myelin which insulates them and increases the speed at which impulses travel. Myelination occurs during fetal development and early childhood and must be quite a drain on the infant's resources. *Homo erectus* may have begun eating more meat than the australopithecines (and making tools to cut it up), primarily to provide for the greater demand of the increasingly greedy brain.

The brain is also a dangerous organ to produce. The fact that large brains came about in a species that was already bipedal may be a coincidence, but it means that we are especially ill-suited to giving birth to our big-brained babies. Various adaptations have made it possible, including the fact that human babies are born extremely premature as compared with most other species. They are helpless and unable to fend for themselves, and are born with soft skulls that only harden up later. A baby's brain is about 385 cc at birth and more than triples in size in the first few years. Even with these adaptations, birth is a hazardous process for modern humans. Many babies and mothers die because the skull is simply too big for an easy birth. All these facts suggest that a powerful and consistent selection pressure for larger brains was at work, but we do not know what it was.

I have so far talked about the increase in brain size as though it were just a simple enlargement, when in fact it is more complicated than that. Higher vertebrates in general have more cerebral cortex than other animals while the older parts of the brain that control breathing, feeding, sleep–waking cycles and emotional responses are more similar. However, the most interesting comparisons are between actual human brains and what might be expected of a typical ape of our size. Although we are highly visual animals our visual cortex (at the very back of the brain) is relatively small while the prefrontal cortex, at the very front, is most enlarged. This difference may well be because our eyes are a normal size and the amount of cortex needed to process the complex visual information coming in, is relatively constant for any ape. The prefrontal cortex, by contrast, does not directly take sensory information but is fed by neurons coming from other parts of the brain.

The frontal cortex is itself a kind of mystery. There is no clear answer to the question 'What does the frontal cortex do?' This is particularly frustrating because if we knew precisely what this part of the brain did, then we might be closer to understanding the selection pressures for the larger brain – but we do not know. People can function surprisingly well with gross damage to this part of the brain, as is known from the famous 1848 case of Phineas Gage, the railroad foreman whose frontal cortex was pierced right through with an iron bar in an explosion. Although his personality was completely changed, and his life and ability to hold down a job were ruined, he could still walk and talk and, at least to some extent, appeared normal. The same is true of the victims of frontal lobotomy, a crude operation that destroyed parts of the frontal cortex and was once used to control serious psychiatric cases. They were never 'themselves' again but the changes were subtle considering the vast amounts of brain damage caused in this horrible 'treatment'. There are numerous theories of the function of the frontal lobes but none is universally accepted. We cannot find out why our large brains evolved by appealing to the function of the part that was enlarged the most.

Apart from the massive increase in the frontal lobes, the brain has been reorganised in other ways. For example, there are two main cortical areas that are critical for language, Broca's area which is responsible for speech production, and Wernicke's area which is responsible for language understanding. Interestingly, these two areas seem to have evolved from the motor cortex and auditory cortex, respectively. Most sounds made by other animals, from grunts to calls and birdsongs are produced in the midbrain, by areas closely connected to those controlling emotional responses and general arousal levels. Some human sounds, such as crying and laughing, are also produced by midbrain areas, but speech is controlled from the cortex. In most people both of the main language areas are in the left hemisphere, so that the two halves of our brains are not the same. Most of us are right-handed, meaning that our left hemisphere is dominant. Although some apes show handedness most do not and there is nothing like our systematic brain asymmetry in other primates. Clearly, our brains have changed in many ways other than just size.

I have described very briefly what needs to be explained – that over a period of about 2.5 million years hominid brains steadily increased in size, an increase that carries obvious costs and must have been driven by a powerful selection pressure. But we do not know what that pressure was.

Theories of the big brain

Theories abound. Most early theories suggested that toolmaking and technological advances drove the need for a larger brain. For theories of this kind the selection pressure came from the physical environment and from other animals. Human brains were needed to outwit their prey. Tools provided obvious advantages and bigger brains could make better tools. Among problems for this kind of theory are that the increase in brain size seems to be out of all proportion to the scale of the endeavour. Big brains are so expensive that if you could catch your prey with a slightly smaller one you would have an advantage. Many pack animals hunt extremely effectively with brains that are small by human standards. Indeed, as we have seen, it rather looks as though *Homo erectus* began eating more meat to feed the growing brain rather than vice versa. Something else must have been driving brain size.

Early hominids obtained much of their food by foraging. So perhaps a big brain was needed for extracting difficult foods or for the spatial ability and cognitive maps needed to find food in patchy and unpredictable environments. However, animals with very small brains manage to store and find food in vast numbers of separate locations, and many animals, such as squirrels and sewer rats, make cognitive maps of large areas. Species with such good spatial skills do show differences in brain structure but not in overall size. Also, predictions concerning brain size and foraging range have not generally supported this kind of theory (Barton and Dunbar 1997; Harvey and Krebs 1990).

Other theories emphasise the social environment. The Cambridge psychologist Nicholas Humphrey (1986) suggested that early hominids took an important step beyond their ancestors by beginning to look into their own minds as a way of predicting what others would do. For example, if you want to know whether that huge male gorilla is likely to attack you if you try to mate with this attractive female, you should try to imagine what you would do in the same situation. This introspection is the origin of what Humphrey calls '*Homo psychologicus*', of humans capable of understanding that others have minds, and ultimately of self-awareness.

Consciousness itself is something we value highly and tend to think of as uniquely human and special, but whether it provides any selective advantage is a fiercely debated issue (e.g. Blakemore and Greenfield 1987; Chalmers 1996; Dennett 1991). Some argue that consciousness could not have evolved unless it had a function, while others maintain that consciousness is not the sort of thing that could have a function. For example, if consciousness is an epiphenomenon of attention or language or intelligence, then the selective advantage would be for those capabilities, not for consciousness itself. More radically, some believe that consciousness is an illusion, or that the whole idea of consciousness will ultimately be dropped, just as the idea of the 'life force' was dropped when we began to understand the mechanisms of life. Clearly, consciousness cannot help us explain the big brain; you cannot solve one mystery by invoking another.

An influential version of social theory is the 'Machiavellian Intelligence' hypothesis (Byrne and Whiten 1988; Whiten and Byrne 1997). Social interactions and relationships are not only complex but also constantly changing and therefore require fast parallel processing (Barton and Dunbar 1997). The similarity with Niccolò Machiavelli (1469–1527), the devious adviser of sixteenth-century Italian princes, is that much of social life is a question of outwitting others, plotting and scheming, entering into alliances and breaking them again. All this requires a lot of brain power to remember who is who, and who has done what to whom, as well as to think up ever more crafty wiles, and to double bluff the crafty wiles of your rivals – leading to a spiralling arms race.

'Arms races' are common in biology, as when predators evolve to run ever faster to catch their faster prey, or parasites evolve to outwit the immune systems of their hosts. The notion that some kind of spiralling or self-catalytic process is involved certainly suits what Christopher Wills (1993) calls 'the runaway brain', and this idea is common among theories that relate language evolution to brain size. These take the social function of the brain a step further but I will leave them until the next chapter. In general, social theories of the evolution of intelligence have been very successful over the past ten years. They have shifted the balance from male-dominated technological explanations to those that appreciate the complexity of social life. Research on the topic is growing but many questions remain. Why, for example, was there pressure for such a great improvement in social skills? Competition within the species is implied but why should just this species and no others take this expensive route? I also wonder how much our peculiar abilities to do mathematics, program computers, paint pictures or build cathedrals actually come down to social skills. Many people think the social theories are the best we have, but the question of brain size is far from resolved. No one knows for sure how and why we got our enormous brains.

Did memes drive brain size?

I am going to propose an entirely new theory based on memetics. In summary it is this. The turning point in our evolutionary history was when we began to imitate each other. From this point on a second replicator, the meme, came into play. Memes changed the environment in which genes were selected, and the direction of change was determined by the outcome of memetic selection. So the selection pressures which produced the massive increase in brain size were initiated and driven by memes.

I shall explore this new theory in two ways, first by rather speculatively looking at our origins again, and then by examining in more detail the processes of memetic driving.

The turning point was when early hominids began to imitate each other. The origins of imitation itself are lost in our far past, but the selective (genetic) advantage of imitation is no mystery. Imitation may be very difficult to do but is certainly a 'good trick' if you can acquire it. If your neighbour has learned something really useful – like which foods to eat and which to avoid, or how to get inside a prickly pine cone – it may pay (in biological terms) to copy him. You can then avoid all the slow and potentially dangerous process of trying out new foods for yourself. This is only worthwhile if the environment does not change too fast, a factor that can be modelled mathematically. Richerson and Boyd (1992) have shown the conditions under which natural selection might favour more reliance on social learning (including imitation) rather than individual learning. Economists have devised models of how optimisers (who incur the cost of making a decision) can coexist with imitators (who avoid the cost but make inferior decisions) and studied how fads and fashions result when large numbers of people all imitate each other (Bikhchandani *et al.* 1992; Conlisk 1980). Indeed, fads and fashions have been associated with imitation ever since Charles Mackay (1841) blamed such 'extraordinary popular delusions' as the South Sea Bubble and the Dutch seventeenth-century mania for tulips on 'popular imitativeness'.

But why did generalised imitation apparently evolve only once? We know from studies of other animals, already discussed, that social learning is fairly common in the animal kingdom but true imitation is very rare. Why should it have arisen in early hominids rather than any other kind of animal?

I suggested that imitation requires three skills: making decisions about what to imitate, complex transformations from one point of view to another, and the production of matching bodily actions. These basic skills, or at least the beginnings of them, are available in many primates and were probably available to our ancestors of 5 million years ago. Primates have good motor control and hand co-ordination, and good general intelligence which would enable them to classify actions and decide what to imitate. Some of them can imagine events and manipulate them mentally, as is shown by their use of insight to solve such problems as reaching food with sticks or by piling up boxes, and, most notably, they have Machiavellian Intelligence and the beginnings of a theory of mind.

The connection between advanced social skills (or Machiavellian Intelligence) and imitation is this. To indulge in deception, pretence and social manipulation you need to be able to put yourself in another's shoes; to take the other's point of view; to imagine what it would be like to be that other. This is precisely what you need to be able to imitate someone else. In both cases you must be able to transform what you see someone else doing into what you would have to do to achieve the same end, and vice versa. Finally, our ancestors used reciprocal altruism, that is, helping others who will later help you in return. As we shall see (Chapter 12), a common strategy in reciprocation is to copy what the other person does; cooperating if they cooperate and refusing if they do not. With all these prior skills imitation is not such a huge leap for evolution to make.

I suggest that the social skills others have singled out as directly responsible for our large brain were in fact responsible for the prior step of acquiring imitation. As soon as our ancestors crossed the threshold into true imitation a second replicator was unwittingly unleashed. Only then did the memetic pressure for increasing brain size begin.

When was this turning point? The first signs of obvious imitation are the stone tools made by *Homo habilis* 2.5 million years ago. We modern people are not natural stone flakers, and experiments trying to find out how it was originally done have shown that making stone tools is a fine art and not easily learned by trial and error. Almost certainly the skill of making stone tools spread through early peoples by imitation. This certainty is much increased in the later archaeological record which shows styles of tools, pots, jewellery and other cultural artefacts spreading through different cultures at different times.

Imitation could have begun earlier. Perhaps people imitated ways of making baskets, wooden scrapers or knives, baby slings, or other useful artefacts that would not have survived the way stone tools do. So let us imagine a very early culture of *H. habilis*, using simple stone tools to cut up and skin game or shape wood, and inventing and copying a few other simple artefacts.

As the new skills begin to spread it becomes more and more important to be able to acquire them. And how do you acquire them? – by imitation of course. Thus, being a good imitator becomes increasingly important. Not only that, but it becomes important to imitate the right people and the right things. In such decisions we would expect simple heuristics, or rules of thumb, to be used. 'Imitate the most successful people' might be one; but now there are memes this does not just mean imitating those with the most food or the strongest muscles, it means imitating the people who have the most impressive tools, the brightest clothes, or the newest skills. What this amounts to is 'imitate the best imitator'. As a consequence, whatever is deemed best spreads fastest.

Another important decision is whom to mate with, and again the answer should be the best imitators, because they will provide you with children who are more likely to be good imitators. All this pressure for better imitation creates more people who are good at spreading memes – whether the memes are ways of making tools, rituals, clothes or whatever. As imitation improves, more new skills are invented and spread, and these in turn create more pressure to be able to copy them. And so it goes on. In a few million years, not only have the memes changed out of all recognition but the genes have been forced into creating brains capable of spreading them – big brains.

That is the story in a nutshell, but I now want to unpack it and take it one step at a time, looking more closely at the mechanisms involved.

The first step we might call 'selection for imitation'. Let us assume, echoing Darwin's original argument, that there is some genetic variation in people's ability to imitate. Some people quickly pick up the new technology of stone flaking, while others do not. Who will do better? The better imitators of course. If stone tools help with food processing, then better toolmakers will eat better and their children will eat better. So far, the same argument could be applied equally well to having strong hands for making the tools. But the difference is this – imitation is a general skill. The good imitators would also be good at copying ways of making wooden scrapers or baskets, or plaiting their hair, or carrying piles of leaves or fruits, or making warm clothes, or any skill that helps survival and can be stolen from someone else. Genes for being a good imitator will begin to spread in the gene pool. Now the environment in which the genes are selected begins to change. If you are absolutely hopeless at imitation, you, and therefore your offspring, will be at a disadvantage in a way you never would have been a few thousand years earlier. The new selection pressure begins with this step.

The next step we might call 'selection for imitating the imitators'. Whom does it pay to imitate? The good imitators of course. Imagine a woman who is especially skilled at copying the latest ways of picking inaccessible fruits or carrying them back to the family, or a man especially good at copying the best toolmaker. If you are an inferior imitator it will still pay you to copy the best imitators. They will have acquired the most useful skills and you now need those skills. During the last millennium you did not. When no one had clothes there was no competitive edge to having them, but now they have been invented you will be less protected from cold and injury and less likely to survive than people who do have them. Now that carrying-baskets have been invented you will get less of the best fruit if you cannot make one. Genes for imitating the best imitators will increase in the gene pool.

Note that this is an escalating process. A male robin can only get a bigger territory in predetermined ways, for example, by singing well – and there is a limit to how brilliant any robin song can be. But a male *Homo erectus* might get power and influence, and come to be copied, by wearing more impressive clothes, lighting bigger and better fires to cook the meat – or to scare the people who have not yet mastered fire – having the sharpest tools, and so on. There is no theoretical limit to this process or to the direction it may take. Selection pressures on the genes will be influenced by whichever memes happen to proliferate. Memes evolve as memes build on memes; new tools emerge; new clothes are made; new ways of doing things are invented. As these memes spread the most successful people are those who can acquire the currently most important memes. Genes for being able to copy the best memes, and genes for copying the people who have the best memes, will be more successful than other genes.

But which are the best memes? 'Best' means, initially at least, 'best for the genes'. People who copy survival-related memes will fare better than people who copy irrelevant memes. But it cannot always be obvious which these memes are. The genes set us up with preferences that reflect their interests. So we like cool drinks and sweet foods, and enjoy sex, for example. These things feel 'best' to us because they were best for the genes of our ancestors. But memes can change faster than human genes, so the genes will not be able to track them effectively. The best the system can do is probably to evolve heuristics such as 'copy the most obvious memes' or 'copy the most popular memes' or 'copy memes to do with food, sex and winning battles'. We will look at the effect of such heuristics in modern society later. In ancient hominid society such heuristics would initially have helped individuals survive and spread their genes, but then increasingly would allow the memes to outwit the genes. Any meme that *looked* popular, sexy, or very obvious would spread in the meme pool and thus change the selection pressures on the genes.

The third step could be called 'selection for mating with the imitators'. In our imaginary society it would pay to mate with the same people you want to copy. If you mate with the best imitators, then your offspring are more likely to be good imitators and so to acquire all the things that have become important in this newly emerging culture. It is this conjunction that drives the process on – first it pays to copy the best imitators because they will have the most useful skills, next it pays to mate with them so that your children can also get these skills. But the heuristics for choosing what to imitate are only rough-and-ready guidelines and the memes are beginning to proliferate beyond purely survival-related skills. For example, as memes for singing appeared, the best imitators would begin to sing best, singing would be perceived as important, and so copying singing would come to have survival value. In this way, the specific nature of the memes of the time would determine which genes were more successful. The memes began to force the hand of the genes.

There is a fourth and final step that might increase the process again – though it is not necessary to the explanation. We can call this 'sexual selection for imitation'. Sexual selection, first described by Darwin and much argued about ever since, is a well-recognised, if controversial, process in biology (see Cronin 1991 for a review and Fisher 1930). Particularly interesting cases involve runaway sexual selection, in which elaborate but otherwise useless structures, such as the peacock's fantastic tail, are selected for by generations of peahens choosing males with fancier tails. Once the process has begun it can incur enormous costs for the male but it works for the following reason. A female who chooses a male with a good tail will have sons with good tails who will attract mates with choices like hers. She will therefore have more grandchildren. The reason it is the females who do the selecting is the imbalance between the sexes in parental investment. Male birds can potentially have vast numbers of offspring but females are constrained to producing only a few eggs a year or, in the case of humans, a few children in a lifetime. So females cannot greatly increase the number of children they have. They can, however, increase the number of their descendants in future generations by choosing mates who will give them 'sexy sons' who will have many offspring. With lots of females all going for the same males, this process rapidly escalates until the costs become too great.

The big brain certainly looks like a runaway phenomenon and I am not the first to suggest a role for sexual selection in brain size. But previous theorists have not explained why sexual selection should pick on brain size (e.g. Deacon 1997; Miller 1993). My answer comes directly from the power of the memes.

The way memes can exploit the process of sexual selection is unique. Whatever is deemed 'in' can change as fast as the memes change – and that is much faster than genes can produce longer tails or an innate ability to build a fancy nest. If you follow the heuristic 'mate with the man with the most memes' you will soon find yourself mating with the one with the best hairdo or the best song (as well as the ability to imitate). If other females start going for good songs then it becomes advantageous to have male children who can pick up a tune quickly. Or if females (for whatever reason) start going for ritualised hunting dances then it becomes advantageous to have male children who can copy dances. The selection pressures on the genes now change in the wake of changes in the memes. The process of sexual selection is exactly the same as it is in examples of biological evolution, but with the added twist that the things being selected for can spread at the speed of memetic evolution. Meme-driven sexual selection will favour mating with males who are not only good at imitating in general, but who are good at imitating whatever happen to be the favoured memes at the time. In this way the memes are, as it were, dragging the genes along. The leash has been reversed and, to mix metaphors, the dog is in the driving seat.

Please note, however, that sexual selection is not necessary for a memetic explanation of brain size, and its role must be an empirical question for the future. The first three processes alone will produce the selection pressures required to drive a runaway increase in brain size – if one further small assumption is made. That is, that being good at imitation requires a big brain. Interestingly, there has been so little attention paid to imitation that there is very little information to back this up. However, this theory suggests that the main tasks of our larger brains are first, the general ability to imitate, and second, the particular ability to imitate the kinds of memes that have proliferated in our species' past.

Can this theory be tested? Like so many biological theories it is not easy to devise specific experimental tests. Nevertheless, some predictions can be made. For example, within any related group of species I would predict that imitation ability will correlate positively with brain size. That is, the best imitators will have the largest brains. Given the scarcity of imitation among other animals there will not be much data to choose from, and there will be problems with choosing an appropriate measure of encephalisation, but this study ought to be possible for various groups of birds and cetaceans.

Using humans, experiments could compare two people performing the same actions but with one person initiating the action while the other imitates it. Various measures could be used to determine just how much extra demand is created by imitating. For example, cognitive studies should show that imitation requires a lot of processing and that we have specialised mechanisms for doing it. Brain-scan studies should show that imitation requires a large amount of energy, and that the extra activity is found predominantly in the evolutionarily newer parts of the brain – those parts that differentiate us from other species. I would not be surprised if specific neurons were found that carry out some of the basic tasks of imitation, such as relating observed facial expressions or actions to one's own, but we will need to know a lot more about how imitation is done before we can guess what to look for.

If these predictions turn out to be right they will confirm the suggestion that imitation is an enormously demanding task, and that it takes a large brain to be able to do it. I would further predict that many aspects of language and thought will turn out to be best understood as by-products of our brains' ability to select which aspects of the world to imitate. However, until more research on imitation is carried out, I can only speculate and say this – if being good at imitation requires a big brain then the processes described above can explain it. These are: selection for imitation, selection for imitating the best imitators, selection for mating with the best imitators, and (possibly) memetic sexual selection. Once early hominids achieved imitation the second replicator was born and these processes began to drive the increase in brain size. The enormous human brain has been created by the memes.

CHAPTER 7

The origins of language

Why do we talk so much?

This might not be a question you have agonised over, but once I started to think about it I found it more and more interesting. How much time and energy does an average person spend on talking every day? I doubt it has been measured but the answer must be several hours. A typical human form of entertainment is to sit over a meal or a few drinks and talk to a lot of other people – what about? Well, about football, or sex, or who has got off with whom, or what he said to her or she said to him, or the latest trouble at work, or the iniquities of the latest government proposals on health care, and so on and on and on. According to some estimates about two-thirds of all conversation is taken up with social matters (Dunbar 1996). It is rare for any group of people to sit in companionable silence.

Then there is work. Some jobs are silent but most are not. In shops and offices, on the buses and trains, in factories and restaurants, people talk. And if they do not talk they often have the radio on with voices and music coming at them from somewhere else. And then there are other forms of communication that use language – the letters, magazines and newspapers arrive on the doormat, the phone rings, the fax starts up, the e-mail messages flood in. The use of time and energy is phenomenal. What is it all for?

There are at least three issues here. One is *why* we talk at all – in other words, why human beings acquired language in the first place. The second is *how* we acquired language – how the human brain became structured the way it did. The third is why, having acquired language, we use it so much. I am going to tackle the last question first, partly because it is easier, and partly because the answer will help us with the much more controversial questions of how and why language evolved.

Why do we talk so much?

Talking all the time must cost energy – and a lot of it. Thinking uses some energy, but talking uses a lot more. Not only are several brain areas necessarily active during speech, or when listening to and understanding speech, but the production of sound is itself expensive. If you have ever been very ill you will know how exhausting it is to speak. You may lie in a hospital bed perfectly able to think but when the nurse arrives you can barely manage a feeble 'thank you', while a few days later you will happily engage in friendly banter about the quality of the food, or what you will do when you get out – complete with smiles, laughs and completely superfluous chit-chat.

Perhaps you are a hi-fi freak. If so you will know how much energy is needed to drive big speakers, and how expensive the sound system gets when it needs to play loud, high-quality sound. Or if you prefer low tech you may have a clockwork radio, in which case you will know all too well, by the feeling in your arm, how much energy is needed to produce that sound, and how much winding you can save by turning down the volume.

This phenomenal use of energy presents something of a puzzle. Living creatures have to work hard for all the energy they consume, and efficient energy use is a critical factor in survival. If you can use less energy than your neighbour, you are more likely to pull through the hard times, to find scarce food, to win the competition for the best mate, and so to pass on your genes. Why, then, has evolution produced creatures that talk whenever they get the chance?

Several possible answers spring to mind. First, there may, after all, be a sound biological explanation. Perhaps talking serves an important function that I have overlooked, such as cementing social bonds or exchanging useful information. I will consider theories of this kind later on.

Second, a sociobiologist might argue that, with the evolution of language, culture has somehow got temporarily out of hand, and the cultural trait of speech has been stretching the leash. However, if talking is really wasteful of precious energy then the genes of the people who talk most will do less well and in time the genes will pull the leash in again.

Third, an evolutionary psychologist might argue that all this talking once had advantages for our ancestors and so we are stuck with it now, even though it doesn't benefit our genes any more. On this view we ought to be able to find the function of so much talking in the lives of early hunter-gatherers.

All these suggestions have in common that they appeal to genetic advantage for an explanation. Memetics provides a totally different approach. Rather than asking what advantage talking provides to the genes, we can ask what advantage it provides to the memes. Now the answer is obvious. Talking spreads memes. In other words, the reason we talk so much is not to benefit our genes, but to spread our memes.

There are several ways of looking at how memes exert pressure on us to keep talking, and I will consider three of them in more detail.

First, since talking is an efficient way of propagating memes, memes that can get themselves spoken will (in general) be copied more often than those that cannot. So these kinds of memes will spread in the meme pool and we will all end up talking a lot.

This argument is similar to the explanation I proposed for why we think so much – another example of the 'weed theory' of memes (p. 41). Silence is like a beautifully weeded flowerbed, just waiting for your favourite plants, and it does not stay that way for long. A silent person is an idle copying machine waiting to be exploited. Your brain is full of ideas, memories, thoughts to be shared, and actions to be carried out; the social world is full of new memes being created, spread about, and competing to be taken up by you and passed on again. But you cannot possibly speak them all. Competition to take charge of your voice is strong – just as competition to grow in the garden is strong. Keeping silence is as hard work as weeding.

So which memes will win in this competition to take over your voice? It may help to ask again our familiar question – *imagine a world full of brains, and far more memes than can possibly find homes. Which memes are more likely to find a safe home and get passed on again?*

Certain memes are particularly easy to say, or almost force their hosts to pass them on. These include bits of juicy scandal, terrifying news, comforting ideas of various sorts, or useful instructions. Some of these have their 'spread me' effect for good biological and psychological reasons. Perhaps they tap into needs for sex, social cohesion, excitement, or avoiding danger. Perhaps people pass them on in order to conform, to be better liked, to enjoy the other person's surprise or laughter. Perhaps the information will be genuinely useful to the other person. We can certainly study all these reasons (and indeed psychologists do just that) but for the memetic argument I am proposing here it does not matter what they are. The point is you are less likely to want to pass on some boring thing you heard about the health of your neighbour's rose bushes than a rumour about what your neighbour was doing behind them. Such 'say me' memes will therefore spread better than other memes and many people will get infected with them.

The news of Princess Diana's death in 1997 spread around the world at the speed of light within minutes of its first announcement. People all over the world told anyone who did not yet know. I did myself. I turned on the radio, heard the continuous coverage instead of the weather forecast and called out to the rest of my family. Then I felt a bit silly for shouting so loud about something I would normally profess to take no interest in. But the death of Diana was just that sort of news. It spread like an extremely infectious virus and within weeks the princess's reputation had become saintly and her following cult-like (Marsden 1997). Within a few months, millions of pounds had been given to her memorial fund and millions more made out of selling her image. Few memes can claim anything like this power, but the principle is quite general. Certain kinds of news spread more effectively than others. These are the things people get to hear about and want to pass on again. As a result, people talk more.

This does not mean that silence is impossible. It is just rare, and needs special rules to enforce it against the natural memetic tendency for endless talk. We see these rules all over the place, in libraries and schools, in lecture theatres and cinemas, and even in special train carriages – and we see people, despite their best intentions, finding themselves breaking the rules. True vows of silence are hard to make, and on religious retreats beginners find the rules of silence difficult to keep, even for a few days. Taking on a silence meme goes against the grain.

This suggests a second approach: to look at rules or social practices concerning speech. Again let's compare two types of meme. Suppose there are instructions encouraging people to talk a lot. These might come in many forms, such as embarrassment at being silent in company, or rules about making polite conversation or entertaining people with chat. Now suppose there are other memes for keeping silent, such as the suggestion that idle chat is pointless, a rule of quiet etiquette, or a spiritual belief in the value of silence. Which will do better? I suggest the first type will. People who hold these memes will talk more; therefore, the things they say will be heard more often and have more chances of being picked up by other people.

If this conclusion does not immediately appear obvious think of it this way – imagine that one hundred people have been taught behaviour of the first type – such as 'You should make polite conservation whenever you can' – and another hundred people have been taught the rule 'It's polite only to talk when you have to'. The first group will, because they hold this meme, talk whenever they have the chance. The second lot will keep quiet. If talkers meet talkers they will all talk. If silents meet silents they will not. The interesting mixture is when talkers meet silent types. It is possible that nobody will ever change their minds or throw out old memes in favour of new ones, but if ever this does happen the imbalance is obvious. A talker will talk, and either directly, or by implication, suggest that polite conversation is necessary, or that talking is fun, or useful. A silent type might be converted. But the reverse is extremely unlikely to happen. The silent type may occasionally say things like 'I think it's better to keep quiet' or 'Why don't you shut up' but will not, by definition, say much – and for that reason alone is unlikely to make converts. Although single memes of this explicitness are probably rare, there are notable examples, such as the British Telecom slogan 'It's good to talk' and the proverbial 'Silence is golden'. Memetics should help us understand not only why talking in general must spread, but also how some selective environments can encourage the rarer silence rule to succeed.

A final way of looking at the memetic pressure to talk is to consider groups of memes or memeplexes, and the kinds of person who will nurture and spread them. Memes that thrive in the environment of a chatty person (and contribute to that person being chatty) will differ from those that thrive in the environment of a silent type. The chatty person will, by definition, talk more and so give her memes more chances of spreading. When another chatty person hears these ideas she will easily pick them up and pass them on again. The silent person will not talk much and so all the memes compatible with being a quiet type will have fewer chances to spread. Of course, chatty people can be extremely irritating, and silent types deeply fascinating, but this does not alter the basic imbalance, the inevitable result of which is that memes for talking, or memes that exist happily with memes for talking, will spread in the meme pool at the expense of memes for keeping silent.

These are several memetic arguments which all conspire to have the same effect. If they are correct, it means that the meme pool gradually fills up with memes that encourage talking. We all come across them and that is why we talk so much. We are driven to talk by our memes.

Memetics thus provides a very simple answer to the question – why do we talk so much. This talking is not for our benefit or to make us happy – though sometimes it may do that – nor is it for the benefit of our genes. It is just an inevitable consequence of having a brain that is capable of imitating speech.

This brings us straight back to our other two major issues – how and why we came to have speech in the first place.

The evolution of language

The question of language origins has been so contentious, that as long ago as 1866 the Société de Linguistique de Paris banned any more speculation about the issue. The glaring gulf between animal communication and human speech cried out for explanation but, with little evidence from palaeontology, the speculations of the time could run wild – our words originated from copying animals or natural sounds, or from making grunts of exertion or disgust. These theories, mockingly dubbed the 'bow-wow', 'ding-dong', 'heave-ho', and 'pooh-pooh' theories, did nothing to explain the origins of grammar and syntax. More than a century later the issue is far from settled and the arguments are still fierce. Our theorising is, however, constrained by a much better understanding of language itself, and by evidence on how the brain and language evolved together.

First, let us look briefly at the nature of modern human language.

Our language capacity is largely innate and not a by-product of intelligence or a general ability to learn – though this was once a hotly debated issue. The fact is that people do not learn language by being systematically corrected for their mistakes, nor by listening attentively and slavishly copying what they hear. Instead, they just seem to pick it up, using minimal input to build up richly structured grammatical speech. Note that by grammar I mean the natural structures of languages that distinguish who did what to whom or when it happened or in what order – not the sort of rule-book grammar that used to be taught at school.

Almost everyone can use language as grammatically as everyone else, regardless of educational attainment or general intelligence. All human societies ever discovered have language, and all of them have complex grammar. Although languages may vary considerably in the size of their vocabularies, they do not differ much in the complexity of their grammar. Hunter-gatherers and remote tribal groups have languages just as complex as modern industrial English or Japanese. Children all over the world can speak grammatically by the time they are three or four years old, and they can invent languages that are more systematic than the utterances that they hear. They can even use subtle grammatical principles for which there is no evidence in the speech they hear. If spoken language is denied them, as for the deaf, they will find other ways of making language. Sign languages are not just simplified or distorted versions of spoken language, but whole new languages that emerge wherever groups of deaf people come together. They are languages in their own right with gestures and facial expressions that take on the grammatical functions of word endings, word orders or inflexion.

This 'language instinct' as Steven Pinker (1994) calls it, singles us out completely from every other species on the planet. As far as we know, no other species has any kind of grammatically structured language – nor are they capable of learning it. When psychologists first tried to teach language to chimpanzees they failed because chimps simply do not have the vocal apparatus to make the necessary sounds. However, they got on better when they trained their chimps with methods that exploit their natural manual dexterity. One chimpanzee, Sarah, has been trained to use a board containing various plastic shapes that represent familiar objects and actions, while Lana and Kanzi press buttons on a special keyboard. Most popular, however, has been the use of signing, building on the fact that chimps have agile hands and make many gestures in the wild. Among the many animals taught this way have been a chimpanzee called Washoe and a gorilla called Koko, both of whom were brought up with humans using American Sign Language.

At first it seemed as though Washoe, Koko and others really could sign. They were credited with 'sentences' three words long, like a child of two years or so. They even made up new words by putting signs together. But the excitement and wild claims soon gave way under careful criticism from psychologists, linguists, and native deaf signers who said that chimp signing was nothing like the rich and expressive human sign language. Wishful thinking probably accounts for much of the exaggeration. The consensus now seems to be that chimps and gorillas can learn single signs or symbols, and use short sequences of them appropriately – mostly to request things. Yet they do not use grammar of any kind and remain oblivious to all the subtleties of sentences that young children seem to take to without effort. Whereas young children just seem to absorb the words they hear and turn them into language, chimps have to be coerced, and rewarded to learn just a few paltry signs. Whatever they may be thinking on the inside (and we should not underestimate that), they just do not 'get' the idea of true language. There is no comparison. It is as though the chimps have to learn the words by the long slow route of ordinary learning – trial and error, and reward and punishment – whereas we just seem to absorb it. The human language capacity is unique.

So how did we get this unique ability? Did it appear all at once in some gigantic lucky leap of sudden evolution (Bickerton 1990)? Or did it appear gradually along with our slowly growing brains? And when did language first appear? Did Lucy indulge in early social chit-chat? Did *Homo habilis* give names to their tools and inventions? Did *Homo erectus* tell stories round their fires?

No one knows for sure. Words do not leave fossils, and extinct languages cannot be brought back. There are, however, a few clues. Some archaeologists believe that we can deduce much about hominid language abilities from their artefacts and burial practices. Only 100 000 years ago there occurred the Upper Palaeolithic Revolution, a time of sudden (in archaeological terms) diversification of hominid activity. For two million years or more the only hominid artefacts had been simple stone tools, the stone flakes probably used as choppers and scrapers by *H. habilis*, and hand-axes made by *H. erectus*. It was not until the Upper Palaeolithic that *H. sapiens* began to leave behind evidence of deliberate burial of the dead, simple painting and body adornment, trading over long distances, increasing sizes of settlements and an extension of toolmaking from stones to bone, clay, antlers and other materials. The view that this dramatic change coincides with the sudden origins of fully developed language is, according to Richard Leakey, common among archaeologists. However, it is based only on speculation. When our own thinking is so bound by the language we learned as children it is almost impossible for us to speculate accurately about what can and cannot be done in the way of art, toolmaking or trading, with what level of language ability. We need better evidence than this.

More solid clues come from anatomy. The major increase in brain size, of roughly 50 per cent, occurred during the transition from the australopithecines to *Homo*. By half a million years ago *H. erectus* had brains nearly as big as ours. Since we do not know the nature of the relationship between brain size and language this cannot tell us when language appeared, but perhaps we can find out something about the organisation of early brains. Obviously brains do not fossilise, but their shape can be deduced from the inside of a fossilised skull. One *H. habilis* skull apparently shows evidence of Broca's area and of the asymmetry characteristic of our language-lateralised brains which led some people to conclude that *H. habilis* could speak. However, recent brain scan studies of living humans show that Broca's area is also active during skilled hand movements and so cannot be definitive evidence for language. Its development might be connected more to the stone tools made by that species. Nicholas Toth of Indiana University has made a detailed study of early stone tools and he and colleagues spent months learning to make them – not an easily acquired skill as it turned out (Toth and Schick 1993). In the process they discovered that most of the early stone tools were made by right-handed people. Brain lateralisation apparently began with the earliest appearance of *Homo* but is not proof of language.

The brain is not the only part of the body that has been modified for speech. Exquisite control of breathing is needed and this meant changes in the muscles of the diaphragm and chest. We have to be able to breathe automatically, as do all land mammals, but then to override the mechanism when speaking, which requires cortical control over the muscles. The larynx is also much lower in humans than in related primates, which makes possible a greater variety of sounds, and the base of the skull is a different shape.

When did these changes take place? Neither larynx nor muscles fossilise, but other clues can be used. One is the base of the skull, the shape of which affects the range of sounds that can be made. It appears flat in australopithecines, slightly flexed in *H. erectus*, and only becomes fully flexed as it is in modern humans, in archaic *H. sapiens*, suggesting that only modern humans could make the full range of sounds that we use now. Another clue comes from the thickness of the spinal cord. Modern humans have much larger thoracic spinal cords than apes or early hominids, presumably because speech requires precise cortical control over breathing. The palaeontologist Alan Walker made a detailed study of a 1.5-million-year-old *H. erectus* skeleton – the 'Nariokotome Boy' found near Lake Turkana in Kenya. This skeleton was well preserved in just the right parts of the spine and showed no thoracic enlargement. In this respect, the Nariokotome Boy was more ape-like than human. As Walker got to know the boy through his ancient remains he became more and more convinced that *erectus* was speechless, and the boy less like a human trapped in an ape's body and more like an ape in a human body. 'He may have been our ancestor, but there was no human consciousness within that human body. He was not one of us,' concluded Walker (Walker and Shipman 1996, p. 235).

All these clues do not give a final answer. Even if we thoroughly understood the anatomical changes involved in producing speech we would not necessarily understand the psychological changes. As psychologist Merlin Donald (1991) points out, there is much more to modern symbolic cultures than language alone, and more than language separates us from our ancestors and from other living primates. Language evolution needs to be understood in relation to the rest of our cognitive evolution.

Perhaps the best we can conclude for now is that language did not appear suddenly, as some linguists have suggested. The evolutionary changes which make modern language possible appear to be strung out over a long period of hominid history. Almost certainly Lucy was incapable of speech, and *H. erectus* could not have had much of a conversation around the fire. Finely controlled speech and fully modern language is unlikely to have appeared until at least the time of archaic *H. sapiens*, little more than 100 000 years ago. That said, the bigger questions remain unanswered. We cannot tell whether the larger brain gradually made language possible, or the beginnings of language gradually forced the increase in brain size. We only know that the two evolved together.

It might help if we knew what language was for.

The answer is not obvious – though it is often portrayed that way. Introductory psychology textbooks tend to make 'obvious' statements like 'The ability to engage in verbal behavior confers decided advantages on our species' (Carlson 1993, p. 271), and leave it at that. The biologists Maynard Smith and Szathmáry (1995, p. 290) start their explanation of language evolution with 'the presumption that natural selection is the only plausible explanation for adaptive design. What other explanation could there be?' Linguists often assume that language 'has an obvious selectional value' or that 'Language must surely confer enormous selectional advantage' (Otero 1990), or talk about language adaptation, the significant selective advantage of communication, or selection pressures for the use of symbols (Deacon 1997).

They are surely right to think in terms of selective advantage. When we ask a 'why' question in biology, the kind of answer we are seeking is usually a functional one. Bats have sonar so that they can catch insects in the gloom. Spiders spin elegant webs to make near invisible, lightweight traps. Fur is for insulation and eyes are for seeing (though the answer never quite stops there!). According to modern Darwinian thinking, all these things gradually evolved because individuals who carried the genes that produced them were more successful at survival and reproduction. If the human language capacity is a biological system like the vertebrate eye or bat sonar then we must be able to say what function it served, and why individuals carrying the genes that increased language competence were more likely to survive and reproduce than their less language-competent neighbours. As we have seen, language cannot have come cheap. Not only are several areas of the brain specialised for understanding and producing speech, but the whole of our vocal apparatus had to be evolved. This meant complex changes in the neck, mouth and throat that compromised other functions; making drinking and breathing at the same time impossible and increasing the risk of choking. Why were these costly and potentially dangerous modifications ever made? What made them worth it?

This question forces us into a difficult situation. As several authors have pointed out (Deacon 1997; Dunbar 1996; Pinker 1994) it appears that either we must understand what selective advantage language gave early hominids, or we must abandon hopes of a Darwinian explanation. This is not a happy choice – if indeed it is a choice.

CHAPTER 8

Meme–gene coevolution

The mystery of language origins has apparently presented us with an unpleasant choice – abandon hopes of a Darwinian explanation or find a function for language. But this is only a forced choice if the function has to be for the genes. If there is a second replicator this is no longer the only option. I shall argue that once imitation evolved and memes appeared, the memes changed the environment in which genes were selected and so forced them to provide better and better meme-spreading apparatus. In other words, the human language capacity has been meme-driven, and the function of language is to spread memes.

What is language for?

If we want to understand the evolution of language, a Darwinian explanation is the obvious starting point. However, it has been argued that language shows no genetic variation, could not exist in intermediate forms, and would require more evolutionary time, and more space in the genome, than could possibly have been available – quite aside from the fact that its selective advantage is not obvious (Pinker and Bloom 1990). All of these arguments have been forcefully opposed. Nevertheless, they keep reappearing in various guises.

Oddly enough, the two major opponents of a traditional Darwinian approach to language origins are one of the world's most famous evolutionary theorists, Stephen Jay Gould, and the world's best-known linguist, Noam Chomsky.

In the 1950s, the prevailing behaviourist paradigm treated language as just another aspect of human beings' general ability to learn. It denied any innate restrictions on what could be learned or any universal properties of language structure. Chomsky went right against this view. He pointed out that the logical structure of languages is far more complex than anyone had thought before, even though it is easily picked up by children without explicit training, and that vastly different languages actually share a common 'deep structure'. He proposed the now familiar idea of an innate Universal Grammar. However, he has subsequently argued that natural selection cannot explain the origin of this Universal Grammar, nor the evolution of language (see Pinker and Bloom 1990). According to Chomsky, we do have innate language structures, but they have not got there by natural selection. They must have got there purely by accident, as a by-product of something else, such as the general increase in intelligence or brain size, or by some other process that we do not yet understand. On this view there were no selection pressures for language itself.

Gould has long argued against the power of selection and adaptation in evolution in general (Gould and Lewontin 1979). He argues, instead, that many biological features evolved as by-products of something else or as consequences of natural physical processes and constraints on structure and form. In the case of language, he says, it must have come about as a by-product of other evolutionary changes – such as the overall increase in brain size (though, as we have seen, that is also unexplained), or because of some as yet unidentified physical constraints.

I do not think that such an approach can work. There is no doubt that simple physical processes can produce intricate designs, such as snowflakes, interference patterns, or ripples on the sand of a beach. There is no doubt that physical constraints are important; the properties of air constrain the shapes of wings and tails, and gravity puts limits on height and size. By-products inevitably occur as designs change, and some of these by-products turn out to be useful and are then exploited by evolution. But these processes alone cannot account for evolutionary progress (though remember that Gould does not believe in progress either) nor for complex functional design. The only process that can produce new designs that build on and develop the old is the evolutionary algorithm (p. 11). With heredity, variation, and selection you can explain the gradual appearance of incredibly improbable things like eyes, ears, fins and tails. Language is an incredibly improbable thing, showing obvious signs of intricate design. It is no explanation at all to say that it came about as a by-product of something else or entirely because of physical constraints.

The non-selectionist arguments of Chomsky, Gould and others have been roundly criticised by Pinker and Bloom and many other contributors to a lively debate in the peer review journal *Brain and Behavioral Sciences* (1990). Pinker and Bloom argue that language shows signs of complex design for some function, and that the only explanation for the origin of organs with complex design is the process of natural selection. They conclude, therefore, that 'specialization for grammar evolved by a conventional neo-Darwinian process' (Pinker and Bloom 1990, p. 707).

But what is the function? A 'conventional neo-Darwinian' explanation assumes a selective advantage of having language. My question about *why* we acquired language now becomes 'what was the selective advantage of having language?' Without an answer to that question the existence of human language remains a mystery.

Pinker and Bloom's answer is that language is designed 'for the communication of propositional structures over a serial channel' (1990, p. 712). But what, then, was the selective advantage of 'communication of propositional structures over a serial channel'? Language would have allowed our ancestors to acquire information and pass it on far faster than biological evolution could achieve, giving them a decisive advantage in competition with other species, they say. But to complete this argument we need to know what biologically relevant information was to be passed on and why the use of propositional structures would have helped. This they do not explain.

There have been many answers before Pinker and Bloom's but none is universally accepted. Some of the earliest theories revolved around hunting. Primitive man was seen as a great hunter who needed to communicate plans for herding prey or trapping them in particular places. In other words, we needed to speak in order to hunt better. A more modern version comes from palaeontologists Walker and Shipman (1996, p. 231) who suggest the function of language was to communicate 'places to hunt; new sorts of traps; locations of water, good caves . . . techniques for making tools . . . or ways to make and keep fire'. Other theories emphasised foraging – perhaps early humans needed to communicate about the locations, nutritional value or safety of available foodstuffs. What is not quite clear from any of these theories is why humans, and humans alone, should have developed such a complex and neurologically expensive solution to the problems of hunting or foraging. For example, wolves and lions achieve clever pack-hunting strategies without grammatical language, and bees communicate the whereabouts and value of food sources with a specialised dance. Vervet monkeys have different warning cries for at least five different predators, including leopard, eagles and snakes (Cheney and Seyfarth 1990) but use no grammar or propositional structures. Presumably, our innate Universal Grammar provides advantages over these simpler systems, but there remains the question why the advantage is so great that we can communicate who did what to whom, why you couldn't make it to the party, and the advantages of the Big Bang theory over a Steady State cosmology.

Perhaps the answer (as in theories relating brain size to Machiavellian

Intelligence, p. 74) lies in the complexity of our social lives. Our hominid ancestors were presumably social animals like their early primate predecessors and, like modern monkeys, we may assume that they could recognise and compare different social relationships and respond appropriately without having verbal labels such as 'friend' or 'sister' (Cheney and Seyfarth 1990). Social primates need to understand matters such as alliances, familial relationships, dominance hierarchies and the trustworthiness of individual members of the group. They also need to communicate. If you are maintaining a complex dominance hierarchy then you need to be able to show (or hide, or pretend to show) fear and aggression, submission and pleasure, desire to be groomed and a willingness to have sex. But emotions are notoriously difficult to talk about. Modern primates get on very well at these complex tasks by means of facial expressions, calls, gestures and other behaviours, and our language does not seem to have been designed to do that job especially well.

The function of language is gossip, says British psychologist Robin Dunbar (1996) – and gossip is a substitute for grooming. He asks the same question I have asked – only rather more poetically 'Why on earth is so much time devoted by so many to the discussion of so little?' In many studies he and his colleagues at Liverpool University have shown that most of our talking is gossip. We discuss each other, who is having what relationship with whom and why; we approve and disapprove, take sides, and generally chat about the social world we live in. Why?

The real function of both grooming and gossip, says Dunbar, is to keep social groups together, and this gets harder and harder as the groups get larger. Many other primates live in social groups and much of their time is taken up with maintaining them. It matters very much who is in an alliance with whom. You chase away your enemies and groom your friends. You share food with your allies and hope they will help you if you are in trouble. You come to your friends' aid – or not, and risk their letting you down next time. Social interactions of this kind demand big brains because so much has to be remembered. You need to remember who did what to whom, when, and how strong or shaky every alliance is at the moment. You will not want to try to steal food from even a low-ranking male if he is in alliance with a stronger one. And you will not risk sex with a receptive female if another stronger male has priority. Also, as group size increases, freeloaders and cheats can more easily escape detection.

How are these complex relationships maintained? For many primates the answer is grooming, but there is a natural limit. As groups get larger the requirement for grooming becomes impossibly high until there are simply not enough hours in the day. Baboons and chimpanzees live in groups of about fifty to fifty-five and spend up to a fifth of their time grooming, but humans live in even bigger groups. We may be able to recognise up to a couple of thousand people, but the more important group size, argues Dunbar, whether in social life, the armed forces or industry, is about one hundred and fifty. Extrapolating from monkeys and apes suggests that we would have to spend an impossible 40 per cent of our time grooming each other to maintain such large groups.

That, says Dunbar, is why we need language. It acts as 'a cheap and ultra-efficient form of grooming' (1996, p. 79). We can talk to more than one person at once, pass on information about cheaters and scoundrels, or tell stories about who makes a reliable friend. So, Dunbar rejects ideas about language being primarily a male-dominated function used for hunting or fighting strategies and suggests instead that it is all about cementing and maintaining our human relationships.

But the obvious question now is why there was selection pressure for larger groups. Dunbar's answer is that our ancestors faced increasing predation as they moved out of the African forests and into the grasslands; safety in numbers would have been a valuable strategy for survival and they had already reached group sizes too large for any more grooming. But many other species have managed other ways of living in the grasslands, some in large herds and others in smaller groups. So could this pressure for larger groups really explain all the drastic and expensive changes required? Dunbar's theory hinges on this point.

Other theories emphasise the evolution of using symbols (e.g. Deacon 1997; Donald 1991). The Harvard neuroscientist Terrence Deacon proclaims humans 'the symbolic species'. He argues that symbolic reference provided the only conceivable selection pressure for the evolution of hominid brains – and by symbolic reference he means the use of arbitrary symbols to stand for something else. Among the advantages of symbolic communication are mother–infant communication, passing on foraging tricks, manipulating competitors, collective warfare and defence, passing on toolmaking skills, and sharing past experiences – 'there are too many compelling options to choose from' he says (p. 377) – but, he argues, these could only have come into play once the 'symbolic threshold' was already crossed. Once true symbolic communication was possible simpler languages (now extinct) would have created a selection pressure for bigger and better brains able to understand them and extend them, leading ultimately to our modern kind of language. But we had to cross 'the symbolic threshold' in the first place.

So how and why did this happen? For marriage, he says. According to Deacon, early hominids could only take advantage of a hunting-provisioning subsistence strategy if they could regulate their reproductive relationships by symbolic means. 'Symbolic culture was a response to a reproductive problem that only symbols could solve: the imperative of representing a social contract' (Deacon 1997, p. 401). On this theory, then, symbolic communication began because it was needed to regulate marriage, and then was gradually improved because of the myriad advantages it provided for other forms of communication.

If I have understood him correctly, Deacon sometimes comes close to a memetic theory. For example, he notes that language is its own prime mover and language evolution a kind of bootstrapping. He even likens one's own language to a personal symbiotic organism. But he does not consider the possibility of a second replicator. For him 'the transmission of genes is the bottom line' (p. 380). Thus, he is stuck with finding the selective advantage *for the genes* of using symbols.

The Canadian psychologist, Merlin Donald, also puts symbolic representation at the heart of his theory (1991, 1993). He suggests that human brains, culture, and cognition all coevolved, passing through three major transitions: mimetic skill, lexical invention (i.e. the creation of words, spoken language and story telling), and finally the externalisation of memory (symbolic art and the technology of writing allowed humans to overcome the limitations of biological memory). His first transition – the development of mimetic skill – sounds as though it might be similar to memetics, but it is not (it is perhaps closer to 'mime' than to 'meme'). Donald clearly distinguishes mimesis from imitation, stressing that mimesis includes representing an event to oneself and is not tied to external communication. He explains: 'mimesis rests on the ability to produce conscious, self-initiated, representational acts that are intentional but not linguistic' (1991, p. 168).

Donald's evolutionary theory, unlike many others, stresses the unique cognitive development of human beings, the importance of their culture, and the consequences of their inventiveness, but he does not invoke the concept of a second replicator. For him, the function of language is part of the wider function of symbolic representation, whose advantage is ultimately to the genes.

I have considered several popular theories of the function of language. All their authors realise there are serious problems, and have tried to explain why language would have given early hominids a selective advantage. I am not convinced that any of them really solves the mystery of human language origins. They need to explain why there is just one species capable of communicating with complex grammatical language, why this one species has a brain so very much bigger than its nearest relatives, and why this one species goes around talking not only about sex, food, and fights, but also about mathematics, the advantages of Macintosh over Windows, and evolutionary biology. There are obviously some advantages to being able to communicate complicated things. When the environment changes, a species that can speak, and pass on new ways of copying, can adapt faster than one that can adapt only by genetic change. Could this be reason enough for all the expensive changes that evolution has brought about in order to give us speech? I do not know. I can only conclude, after this necessarily brief review of the existing theories, that there is no real consensus over the issue.

The situation can be summarised like this. Darwinian accounts of the evolution of human language have assumed that language provided a selective advantage to the genes, but despite many suggestions there is no unanimous agreement on what this selective advantage was. However, this argument assumes that Darwinian explanations must rest entirely on genetic advantage. If we add a second replicator the argument changes completely.

Language spreads memes

Memetics provides a new approach to the evolution of language in which we apply Darwinian thinking to two replicators, not one. On this theory, memetic selection, as well as genetic selection, does the work of creating language. In summary, the theory is this. The human language faculty primarily provided a selective advantage to memes, not genes. The memes then changed the environment in which the genes were selected, and so forced them to build better and better meme-spreading apparatus. In other words, the function of language is to spread memes.

This is a strong claim and I shall therefore take the argument slowly, building on our understanding of coevolution.

I have already explained how meme–gene coevolution could have produced the big brain. To summarise – once imitation has evolved, a second replicator comes into being which spreads much faster than the first. Because the skills that are initially copied are biologically useful, it pays individuals both to copy and to mate with the best imitators. This conjunction means that successful memes begin to dictate which genes are most successful: the genes responsible for improving the spread of those memes. The genes could not have predicted the effect of creating a second replicator and cannot, as it were, take it back. They are now driven by the memes. This is the origin of the dramatic increase in brain size. This theory predicts not only an increasingly large brain but a brain that is specifically designed to be good at spreading the most successful kinds of memes. I shall argue that this is exactly what we have, and that this explains the evolution of language.

If successful memes drive the evolution of the brain, then we need to ask which memes these are. To some extent the success of memes is a matter of serendipity and accidents of history. In our long past it might have been the case that long hair or ringlets, painted faces or scarred legs, singing, worshipping the sun or drawing pictures of insects, came to be the favoured memes. These would then have exerted pressure on the genes to provide brains that were especially good at copying these particular things. If the forces of accident were the major pressures in memetic evolution we would have little hope of ever making sense of our past. However, I am going to assume that overwhelming these forces of serendipity are the fundamental principles of evolutionary theory. That is, there are some basic qualities that make for a successful replicator – in this case a meme.

Dawkins (1976) identifies three criteria for a successful replicator: fidelity, fecundity, and longevity. In other words, a good replicator must be copied accurately, many copies must be made, and the copies must last a long time – although there may be trade-offs between the three. We must always be careful of comparisons with genes, but we can usefully consider how they match up to these requirements.

Genes are high on all three. Their method of replication is extremely accurate. That is, genes have high fidelity in the sense that very few errors are made when long sequences of genetic information are copied. When errors are made there are elaborate chemical systems for repairing them. Of course, there are some remaining errors, and these contribute to the variation that is essential for evolution, but the errors are very few. Also, the process is digital, which makes for much higher fidelity, as we have already seen.

Genes, at least some of them, are extremely fecund, producing masses and masses of copies, though the fecundity varies with the kind of environment a species inhabits. Biologists distinguish two kinds of selection at the extremes of a continuum: *r*-selection and *K*-selection. *r*-Selection applies in unstable and unpredictable environments where it pays to be able to reproduce rapidly and opportunistically when resources allow. High fecundity, small size and long-distance dispersal are favoured, as in frogs, flies and rabbits. *K*-Selection operates in stable, predictable environments where there is heavy competition for limited resources. Such conditions favour large size, long life and small numbers of well-cared-for offspring. *K*-Selected species include elephants and humans. These are the extremes, but even in the most *K*-selected species many copies of the genes are made.

Finally, genes are long lived. Individual molecules of DNA are well protected inside cells, and those that are passed down through the germ line can sometimes survive as long as the lifetime of the organism. Depending on the size of unit you count as the gene, its lifetime varies, but in some sense genes are immortal, since they are passed on from generation to generation to generation. Genes are extremely high-quality replicators.

Were they always that way? Presumably not, although we do not know much about the early history of DNA. However, it is reasonable to assume that the first replicators were simpler chemicals than present-day DNA, and were not packaged efficiently in chromosomes inside cell nuclei and with a complex cellular machinery devoted to their maintenance and duplication. They may, for example, have been simple autocatalytic systems that give rise to two identical molecules, followed by polynucleotide-like molecules, and then RNA (Maynard Smith and Szathmáry 1995). But why should these chemicals have evolved to produce the high-quality replicating system that we have today?

Imagine the competition between various forms of early replicator in their primeval soup. If a low-fidelity replicator and a high-fidelity replicator existed at the same time, the high-fidelity one would win out. As Dennett (1995) puts it, successful evolution is all about the discovery of 'good tricks'. A replicator that makes too many mistakes in copying soon loses any good tricks that it stumbles across. A high-fidelity replicator would not stumble upon them any quicker (and arguably could be slower) but at least it would keep any it found – and thus outperform the competition. Similarly a highly fecund replicator would, simply by virtue of making more copies, swamp its rivals. Finally, a long-lasting replicator would still be around when its competitors had fizzled out. It is obvious really. In this early environment there would have been selection pressure for better and better replicators, and this could ultimately have resulted in the exquisite cellular machinery for copying DNA.

The same principle can be applied to memes. Imagine early hominids who have discovered the biologically 'good trick' of imitation. Initially, this good trick allowed some individuals to profit by stealing the discoveries of others, and these individuals therefore passed on the genes that made them imitators until imitation became widespread.

Then a new replicator was born and, using the copying machine of the brain, began to make copies – copies of actions, copies of behaviours, copies of gestures and facial expressions, and copies of sounds. This world of early memes is the memetic equivalent of the primeval soup. Which of these potentially copyable actions will be more successful as a replicator? The answer is those with high fidelity, high fecundity, and longevity.

Now we can see the relevance of language. Language certainly improves meme fecundity. How many copies of an action can you spread at once? As many copies as there are people watching. But not many people can watch one person perform at once, and also the people nearby may just not be looking, or may get bored and look at something else. On the other hand, if you make a sound, many people can potentially hear it at once and they do not need to be looking – they can even hear it in the dark. This advantage is obvious in the difference between sign languages and speech. They may both be effective for private conversations but you cannot shout to the masses 'Hey, you must listen to this' in sign language. The masses have to be looking first. Also, sound can travel over fairly long distances and round corners. A lot more copies can be made by shouting out your news than by demonstrating it with hand signs, facial expressions, bodily movements or any of the other available signals.

This means that vocalisation is a good candidate for increasing fecundity and thereby winning the battle to become the better replicator. How, then, could the fidelity of the copies of the sounds be increased? One obvious strategy is to make the sounds digital. As we have seen, digital copying is far more accurate than analogue, and genes have certainly adopted the 'get digital' strategy. I suggest that language has done the same. By making discrete words instead of a continuum of sound, copying becomes more accurate.

We might imagine many versions of early verbal language going on simultaneously as people began imitating each other. Any which divided speech up into discrete, easily copyable sounds would have higher fidelity and hence outperform the others in the race to get copied. The problem with copying always lies in deciding which aspects of the stimulus are the important ones to copy. Language is a system that makes these decisions clear by, for example, breaking up the sounds, and adopting norms of pronunciation, while ignoring overall pitch. Note that other forms of communication, such as the warning cries of monkeys, can become gradually more and more distinct by genetic selection, but the process described here works much faster because it spreads from person to person within one generation. Because higher fidelity copies spread more effectively, they tend to predominate, and the language improves.

What about longevity? No individual behaviour has much longevity in itself, but longevity inside the brain is important. Some actions are hard to remember and therefore hard to copy, especially after a delay. We would expect the successful memes to depend on behaviours that are easily remembered so that they can be reproduced even after long delays. Language has very efficiently improved memorability, remembering dance steps can be troublesome, but remembering 'slow, slow, quick-quick slow' is easy. We find it impossible to reproduce a long series of meaningless noises, but easy to repeat back a sentence of a few dozen words. Without too much trouble we can repeat whole stories and conversations. Indeed, many cultures have depended entirely on rote learning of long stories and myths to pass on their history. By structuring the meanings of sounds, language makes them far more memorable.

We can look to technology for another kind of longevity – as when the invention of pots creates long-lasting models for new pots and more pot-making behaviour, or when the building of bridges spreads the idea of a bridge to everyone who crosses one. The longevity of language took a dramatic turn with the invention of writing – committing words to clay, papyrus or floppy disk – but I shall consider these further steps in longevity later.

I have described the appearance of words as a process of digitising. The real problem for understanding language origins is not so much the words, which at least in principle can be learned by simple associative learning, but the grammar. However, grammar also improves replication. How many things can you say with a given set of words? Not very many, unless you have some way of specifying different meanings if you combine the words in different ways. Adding prefixes and suffixes, inflecting them in different ways, and specifying word, order would all increase the number of possible separate utterances that could be produced and copied. In this sense, grammar might be seen as a new way of increasing fecundity as well as fidelity. The more precisely the copies are made, the more effective they will be. Then, as more and more possible things can be said, more memes can be created to continue driving the process.

Remember that all that is going on here is selection, with no need for conscious foresight or deliberate design on the part of either the memes themselves or the people who are copying them. We need only imagine groups of people who all tend to copy each other, and they copy some sounds more than others. Whether a particular sound is copied because it is easy to remember, easy to produce, conveys a pleasant emotion, or provides useful information, does not matter as much as the general principle, that when lots of sounds are in competition to get copied, the successful ones will be those of high fidelity, high fecundity, and longevity. This is the selection pressure that produced grammatical language.

The development of language was thus an evolutionary process like any other, creating complex design apparently out of nowhere. The early products of human sound copying changed the environment of memetic selection so that more complex sounds could find a niche. Just as multicellular organisms could arise only when single cells were already common, just as animals could appear only when plants were already producing oxygen, just as large predators could evolve only when there was plenty of small game about, so complex grammatically structured utterances could appear only when simpler ones were already common. A language with lots of words and well-defined structures would seem to be the natural result of memetic selection.

The next step is to understand how language itself was able to restructure the human brain and vocal system for its own propagation. This is meme–gene coevolution again and works as follows. I have assumed that people will both preferentially copy and preferentially mate with the people with the best memes – in this case the best language. These people then pass on *genetically* whatever it was about their brains that made them good at copying these particularly successful sounds. In this way, brains gradually become better and better able to make just these sounds. Grammatical language is not the direct result of any biological necessity, but of the way the memes changed the environment of genetic selection by increasing their own fidelity, fecundity, and longevity.

Note that this whole process is self-sustaining. Once language evolution begins, both the language itself and the brain on which it runs will continue to evolve under the combined pressure of memetic and genetic selection. This is not the only theory to treat language as 'its own prime mover', or as a self-sustaining process, but others have trouble with explaining how it ever began or why it takes the form it does. Deacon, for example, had to find a reason for crossing the 'symbolic threshold' in the first place. There is no such problem with the memetic theory of language origins. The critical step was the beginning of imitation – and there is no mystery about why natural selection would have favoured imitation. It is an obvious, if difficult to find, 'good trick', and one that is especially likely to arise in a species that already has good memory and problem-solving skills, reciprocal altruism, Machiavellian Intelligence and a complex social life. Once found, it sets in motion the evolution of a new replicator and its coevolution with the old.

I have done a lot of speculating and imagining here. Am I just making another equivalent of the 'bow-wow' or 'heave-ho' theories? Should I be reminded of the ban made by the Société de Linguistique de Paris?

I hope not. The difference here is that I am not suggesting that words arose because people heaving on heavy rocks made 'heave-ho' noises and began to speak – though I suppose the odd word might have come about that way. I am suggesting that verbal language is almost an inevitable consequence of memetic selection. First, sounds are a good candidate for high-fecundity transmission of behaviour. Second, words are an obvious way to digitise the process and so increase its fidelity. Third, grammar is a next step for increasing fidelity and fecundity yet again, and all of these will aid memorability and hence longevity. Once the second replicator arose, language was more or less inevitable.

The theory depends on a few basic assumptions, and these could be tested. One is that people preferentially copy the most articulate people. Social-psychological experiments show that people are more easily persuaded by 'good talkers' and 'fast talkers', but this needs more systematic research using tests of imitation.

Meme–gene coevolution assumes that people preferentially mated with the best meme-spreaders, in this case the most articulate people. We should remember that past selection for 'good talkers' may have used up most of the original variation, leaving most of us fairly articulate today. However, the preference may still be there, so that being highly articulate makes you sexually attractive. The history of love poems and love songs suggests as much, as does the sexual behaviour of politicians, writers and television stars (Miller 1993).

If the theory is right then human grammar should show signs of having been designed for transmitting memes with high fecundity, fidelity, and longevity, rather than to convey information about some particular topic such as hunting, foraging or the symbolic representation of social contracts. This is the memetic equivalent of adaptationist thinking in biology and I might be criticised for assuming that memetic evolution must always have found the best solution and for a kind of circular reasoning. Nevertheless, adaptationist thinking has been extremely effective in biology and may prove so in memetics.

Language continually evolves, and new words or expressions compete to be adopted, or co-opted from other languages. Again, we should expect the winners to be those of high fidelity, fecundity, and longevity. Wright (1998) has used memetics to study the translation of chemical terms such as acid, alcohol, or various elements, into Chinese, showing that alternative terms underwent intense competition for survival, with the winners depending both on properties of the terms themselves and on the meme products already in existence at the time.

Whole languages also compete with one another for survival. Where languages have coexisted in the past we would expect the survivor to be the better replicator, and that languages with especially low-quality replication would most easily be destroyed. Now that so many languages are threatened with extinction this memetic approach might help us to understand what is happening. There is also a battle waging between the major world languages for dominance (or just survival) in industry, finance, transport, and information technology. Historical accidents have made some better placed than others, but we might usefully look at the evolution, competition, and extinction of languages with three things in mind – the fidelity, fecundity, and longevity of the memes they convey.

Finally, we should be able to predict how artificial languages could arise. There have been many attempts to get robots, or virtual robots, to use language. These usually begin by teaching the artificial systems a lot about natural languages, or by getting them to make associations between sounds and objects. The theory I have proposed suggests an entirely different approach that assumes no knowledge of any prior language, and no concept of symbolic reference.

Let's imagine a group of simple robots, ambling about in some kind of relatively interesting and changing environment. We can call them copybots. Each copybot has a sensory system, a system for making variable sounds (perhaps dependent on its own position or some aspect of its sensory input), and a memory for the sounds it hears. Most importantly, it can imitate (though imperfectly) the sounds it hears. Now, imagine that all the copybots start roaming around squeaking and bleeping, and copying each other's squeaks and bleeps.

The environment will soon become full of noise and the copybots will be unable to copy every sound they hear. Depending on how their perception and imitation systems work, they will inevitably ignore some sounds and imitate others. Everything is then in place for the evolutionary algorithm to run – there is heredity, variation, and selection – the sounds (or the stored instructions for making the sounds) are the replicator. What will happen now? Will there just be an awful cacophony, or will something interesting emerge? If the theory is correct then some sounds will have higher fidelity, longevity, and fecundity (depending on characteristics of the copybots) and these should be copied more and more accurately, and patterns begin to appear. Some sounds would be made more often, depending on events in the environment and the positions of the copybots themselves. I think this could be called language. If so, it would not be the same as any language currently used by any natural or artificial systems.

If this worked, interesting questions would arise. Are the copybots really communicating? Are they talking *about* something? If so, symbolic reference would have arisen out of simply providing the robots with the capacity to imitate. In other words, the capacity to imitate is fundamental, not the capacity for symbolic reference. That is exactly what I would expect. The final question is, could we ever understand them?

To summarise, there is a memetic solution to the mystery of human language origins. Once imitation evolved, something like two and a half or three million years ago, a second replicator, the meme, was born. As people began to copy each other the highest-quality memes did the best – that is those with high fidelity, fecundity, and longevity. A spoken grammatical language resulted from the success of copyable sounds that were high in all three. The early speakers of this language not only copied the best speakers in their society but also mated with them, creating natural selection pressures on the genes to produce brains that were ever better and better at spreading the new memes. In this way, the memes and genes coevolved to produce just one species with the extraordinary properties of a large brain and language. The only essential step to starting this process was the beginning of imitation. The general principles of evolution are enough to account for the rest.

The answers to two difficult questions are now obvious, and the same. What is the big brain for? What is the function of language? – To spread memes.

CHAPTER 9

The limits of sociobiology

I have proposed two new theories – memetic theories – to account for human brain size and the origins of language. They both depend on the replicator power of the meme, and introduce some new principles into the way memes and genes interact – the processes I have called 'meme–gene coevolution' and 'memetic driving'. I want now to set this memetic approach in context; to see how it compares with other theories and explain why theories based purely on biological advantage must fail. By exploring the different ways in which memes and genes can interact we will come up against the limits of sociobiology.

First, theories of 'coevolution' are not new. As I explained in Chapter 3 there have been many, including those of Boyd and Richerson (1985), Deacon (1997), Donald (1991), Durham (1991) and Lumsden and Wilson (1981). What makes the present theory of meme–gene coevolution different is that both halves – the memes and the genes – are replicators in their own right, with equivalent status. Certainly, the two replicators are different. They differ in how they work, how they are copied, and the timescales over which they operate. There is also an important asymmetry between them in that memes can operate only by using the brains created by genes, whereas genes can (and do) operate perfectly well without memes. Nevertheless, both memes and genes have replicator power. They are essentially only out for themselves and if they can get copied they will – the rest follows from there.

Dawkins complained that his colleagues always wanted to go back to biological advantage. This theory does not go back only to biological advantage, but to memetic advantage as well. With two replicators working together things can get complicated, but not impossibly so, and with a bit of simplification we can tease out the three major types of interaction: gene–gene interactions, gene–meme interactions, and meme–meme interactions.

Gene–gene interactions

Gene–gene interactions are the stuff of biology. When white bears manage to stalk more seals on the arctic ice than brown bears do, genes for producing white fur spread at the expense of genes for producing brown fur. In this way, rival versions of genes (alleles) compete with each other. Genes also cooperate, however – otherwise we would not have organisms at all. In our own bodies, thousands of genes cooperate to produce muscles and nerves, liver and brain, and to result in a machine that effectively carries all the genes around inside it. Gene–gene cooperation means that genes for digesting meat cooperate with genes for hunting behaviour, while genes for digesting grass cooperate with genes for grazing and chewing the cud. Of course, they do not cooperate out of kindness but because it benefits their own replication to do so.

But these are not the only kinds of gene–gene interaction. Genes in one creature can affect genes in another. Mouse genes for fast running drive cat genes for pouncing quicker. Butterfly genes for camouflage drive better eyesight in birds. In this way, 'arms races' develop in which each creature tries to outwit the other. Many of the most beautiful creations of the natural world are the result of genetic arms races. Organisms exploit each other, as when ivy uses a tree to get height without building its own trunk, or parasites live inside the bodies of people and get their food for free. But others cooperate with each other in symbiotic relationships, such as ants and aphids that provide each other with protection and nourishment, or the many bacteria that live inside our own intestines and without which we could not digest certain kinds of food. It is even thought that the tiny mitochondria that provide the energy inside every living cell originated from symbiotic bacteria. They have their own genes, and these mitochondrial genes are passed on from mother to child in addition to all the more familiar human genes in the cell nucleus.

Another way of looking at the world is to see whole ecosystems as constructed by the interactions between selfish genes. Genes can have multiple effects (a single gene *for* a single effect is a rarity), and may be packaged inside different organisms. Dawkins (1982) provides many examples of what he calls the 'extended phenotype', by which he means all the effects of a gene on the world, not just on the organism in which it sits. Beavers build dams and those dams are as much an effect of genes as are spiders' webs or snail shells or human bones. But the genes concerned need not even be part of the organism that builds the structure. For example, there is a parasitic fluke that lives inside snails and causes them to grow thicker shells. Dawkins suggests that the thickness of a snail's shell is a trade-off between growing a thick shell to protect it from birds, and saving the resources to make more baby snails. The fluke genes will not benefit from more baby snails, but they will benefit from a safe snail to live in – so fluke genes for making snails grow thicker shells are good replicators. This illustrates the important point that although the interests of a gene and the interests of the organism it sits in usually coincide, they do not always do so.

These few examples show how genes (without foresight or intention, and just because they may either be successfully copied or not) can compete with each other, exploit each other, or cooperate with each other for mutual benefit. We can see not only the complexity of gene–gene interactions, but why it is helpful to look at the world from a gene's eye view. None of this makes as much sense if you concentrate only on the individual organisms, even though they are the vehicles that ultimately live or die. The whole complex system is better viewed as driven by the interplay between selfish replicators – in this case genes.

I am going to apply exactly the same principles to meme–meme interactions later – and these will prove to be just as intricately complex. Meme–meme interactions are the stuff of today's society; of religion, politics, and sex; of big business, the global economy, and the Internet. But that comes later. First, we need to clarify the interactions between genes and memes – that is, meme–gene coevolution.

Meme–gene interactions

When memes interact with genes we might expect to find both competition and cooperation, and every gradation in between. As we have seen, several theorists have likened memes to symbionts, mutualists, commensals or parasites. The first was Cloak who said that at best we are in symbiosis with our cultural instructions. 'At worst, we are their slaves' (Cloak 1975, p. 172). Delius (1989) suggests it started out the other way around. The memes were originally the slaves of the genes but, as he says, slaves have a well-known bent towards independence and now our memes may be anything from helpful mutualists to destructive parasites (see also Ball 1984). And Dawkins famously treats religions as viruses of the mind. All this raises the question of whether the memes are the friends of the genes or their enemies.

The answer is, of course, both. But for the sake of sorting out meme–gene interactions I want to divide the interactions into two categories:

those in which the genes drive the memes and those in which the memes drive the genes. This is an oversimplification in many ways. You can imagine cases in which the two help each other equally and no driving really takes place, but more commonly, I suggest, there is at least some imbalance and one replicator or the other predominates.

The reason for this crude distinction is this. When the genes are doing the driving (and the dog is safely on its leash) we have all the familiar results of sociobiology and evolutionary psychology. The interests of the genes predominate and people behave in ways which, somehow or other, give them (or would have given their ancestors) a biological advantage. Men are sexually attracted to women who appear to be fertile; women are attracted to strong, high-status men; we like sweet foods and dislike snakes; and so on (see e.g. Pinker 1997). These effects are very powerful in our lives, and we should not underestimate them, but they are the stuff of biology, ethology, sociobiology and evolutionary psychology – not memetics.

When the memes are doing the driving (and the dog is in charge) power shifts towards the interests of the memes and the results are rather different. These are results that cannot be predicted on the basis of biological advantage alone, and they are therefore critical for memetics. They are what distinguishes memetic theories from all others and are likely, therefore, to be a major testing ground of the value and power of memetics as a science.

I have given two examples of memetic driving so far: the big brain and the origins of language. I shall return to those and add more later, but first let us briefly consider the claim of sociobiology and evolutionary psychology to be able to account for human behaviour and human culture.

Overthrowing the Standard Social Science Model

The argument is exemplified by John Tooby and Leda Cosmides, from the University of California, who plead for a new approach to the psychological foundations of culture (Tooby and Cosmides 1992). They describe the old approach as the Standard Social Science Model, a model that treats the human mind as an infinitely flexible blank slate that is capable of learning any kind of culture at all and is almost entirely independent of biology and genes. Quite rightly (in my opinion) they, and others, have undermined the central assumption of the SSSM.

First, the human mind is simply not a blank slate. In particular, work in artificial intelligence has proved that it could not be so, because an all-purpose general perception machine just cannot get around in the world. To live and feed and reproduce at all it is essential to be able to see objects, track them, grasp them, identify individuals, discriminate between the sexes, and so on. None of this can be done at all without mechanisms for dividing up the world in relevant ways. The world itself can potentially be divided up in an infinite number of ways. Our brains must have, and do have, ways of limiting this potential. They have object-recognition modules, colour-perception systems, grammar modules, and so on (Pinker 1997). The way we experience the world is not 'the way it really is' but the way that has proved useful to natural selection for us to perceive it.

Similarly, learning is not some all-purpose general ability that starts from scratch. Even where imitation is concerned this has proved to be true. In the 1940s and 1950s, when learning theory was being applied to almost every aspect of behaviour, psychologists assumed that imitation itself must be learned by being rewarded. They strongly denied any claims of an 'imitation instinct' and poked fun at older theories of human instinctive behaviour (Miller and Dollard 1941). This was, in the circumstances, understandable. These early theories had included instincts such as a girl's instinct to pat and tidy her hair or, when thrown a ball while sitting, to part her legs and catch it in her skirt. Nevertheless, they were wrong about imitation. Recent research shows that babies begin to imitate facial expressions and gestures from an early age whether they are rewarded or not. Babies are able to mimic facial expressions they see and sounds they hear when they are too young to have learned by practice or by looking in mirrors (Meltzoff 1990). Successfully imitating something seems to be rewarding in itself. We can see now, as the behaviourists could not, why so much of our behaviour has to be instinctive. The world is too complicated to cope with if we have to learn everything from scratch. Indeed, learning itself cannot get off the ground without inbuilt competencies. We humans have more instincts than other species, not fewer. As Steven Pinker puts it 'complexity in the mind is not caused by learning; learning is caused by complexity in the mind' (1994, p. 125).

The old SSSM is clearly being overthrown by the evidence, as some delightful examples can show. One concerns the naming of colours. Anthropologists, working in the old SSSM mode, had long taken colour naming to be a perfect example of cultural relativity. Lots of languages had been studied and wide variation found in the words used to describe colours. In the early 1950s, for example, Verne Ray gave colour samples to sixty native American Indian groups and asked them to name them. He concluded that there is no such thing as a 'natural' division of the spectrum but that each culture has taken the spectral continuum and has divided it on a completely arbitrary basis. In other words, all the colours we call green might be divided into two or more other categories in a second language, combined with some other colour in a third, overlap with different ones in a fourth and so on. This is a strange thought. For us, the experience of seeing red is quite different from that of yellow. We know, when we look at the spectrum, that yellow forms only a thin band between the red and green parts, and this yellow really stands out as being different. It is hard to imagine that another culture would divide this obvious looking spectrum in a totally different way. Yet, this is what the relativity hypothesis implied – that our experience of colour is determined by the language we have learned – either that, or there must be a lot of people in the world who experience sharp divisions between the colours they see but have learned to use names based on quite different divisions.

This view was accepted more or less without question until, many years later, two other anthropologists set out to extend and reconfirm the findings. Brent Berlin and Paul Kay (1969) used a wider range of languages and a more systematic set of colour samples – and they failed. What they found instead was an extraordinarily systematic use of colour names in language after language, and moreover, one that makes sense in terms of the physiology of colour vision. In the visual system, information about brightness is coded separately from colour information. Colour information from three kinds of receptor in the eye is fed into an opponent system that codes colours on one red-green dimension and another yellow-blue dimension. Berlin and Kay found that all languages contain terms for black and white. If a language only has three terms then the third is for red. If it has four terms then the next one is either green or yellow and if it has five then it has both green and yellow. If a language has six colour terms then it includes blue and if seven it includes brown. Languages with more terms then add purple, pink, orange, grey, and so on. Colour naming is not arbitrary and relative, it reflects very well the way our eyes and visual systems have evolved to make use of relevant information in the world around us.

Colour naming has been a favourite for stories of this kind. Have you ever heard that the Eskimos have fifty words for snow? You might even have read that it is more than a hundred, or two hundred or four hundred. None of these is true. Indeed, the 'Great Eskimo Vocabulary Hoax' is a kind of urban myth, an extremely successful meme that has been printed, reprinted, broadcast and spread in numerous other ways, despite being false. Apparently, in 1911 the famous anthropologist Franz Boas noted that Eskimos used four unrelated words for snow. Somehow this idea appealed and was inflated again and again until it became hundreds. Modern estimates suggest that Eskimos use at most a dozen snow words but then this is not many more than there are in English, and is not a bit surprising since Eskimos spend their lives in the snow. Even in English, we have hail, sleet, slush and wintry showers, and people who work in snow, or ski, use extra words as required, such as corn, spring, or sugar snow, powder or (as my dad used to call the wet heavy stuff) puddin'.

The legacy of Boas and extreme cultural relativism stretches far beyond the small matter of the number of words for frozen rain. According to the relativists just about every aspect of human behaviour was learned, was variable, and could be entirely different in different cultures – even sexual behaviour.

Many people seem to hate the idea that human sexuality can be explained in terms of genetic advantage, and early sociobiologists were pilloried for suggesting it. The popular view had long been that familiar sex differences, such as female choosiness and male promiscuity, were purely cultural creations, and in another culture things might be totally different. Superficially, this is certainly true in that some cultures prize huge feather headdresses and others pinstripe suits; some cultures admire naked pendulous breasts and others uplift bras. But what about the more basic differences? The view that all sexual behaviour is culturally determined was central to the work of Franz Boas, and in the 1920s a young student of his, Margaret Mead, set out for Samoa to study a society that he and she believed would be totally different from their own. In her famous book *Coming of Age in Samoa*, Mead (1928) described an apparently idyllic and peaceful life in which there were no sexual inhibitions, and adolescent girls were free to have sex with whomever they liked. Culture, it seemed, was to blame for our own inhibitions and unfair disparities between the sexes. Biology was irrelevant.

This view apparently fitted with what people wished to believe about their own sexual nature, and was accepted as valid evidence that in other cultures almost anything goes. It was a set of successful memes that endured for nearly sixty years even though it was based on only a brief study by a young student. The principle was barely questioned and almost nobody bothered to check. That is until the early 1980s when an Australian anthropologist, Derek Freeman, painstakingly took the story to pieces.

Freeman (1996) spent six years in Samoa and, unlike Mead who was there only four months, he lived with Samoans and had time to learn their language. What he found could hardly have been more at variance with Mead's descriptions of Samoan life. He found aggressive behaviour and frequent warfare, severe forms of punishment for misdemeanours, high rates of delinquency among adolescents and, most important for Mead's thesis, that the Samoans placed great value on virginity. They even had a virginity test and ceremonial defloration of girls at their wedding.

How could Mead have got it so wrong? Freeman was able to track down some of her original informants and found out. One woman, by then aged 86, explained that Mead didn't realise they were 'just joking' when they said they went out at nights with boys. Another confirmed that they had dreamed up the stories for fun – and just imagine the fun of inventing wild and crazy stories about your sex life for an ignorant young visitor who is anxiously writing it all down. As is so often the case, it took a lot more time and hard work to unmake the myth than to make it in the first place. It also took a lot of courage. Freeman's discoveries were scorned by people who had almost made Mead into a guru, and he was vilified for daring to suggest she was completely wrong.

Looking back with the benefit of modern evolutionary psychology we can see how and why the original theories were completely wrong. Cosmides and Tooby are right to reject them. However, their version of evolutiontary psychology seems to me to go too far in the opposite direction. They leave no room for any true evolution of culture. As far as they are concerned 'Human minds, human behavior, human artefacts, and human culture are all biological phenomena' (Tooby and Cosmides 1992, p. 21). In other words, the world of ideas, technology and toys, philosophy and science are all to be explained as the products of biology – of evolution by the natural selection of genes.

I do not wish to underestimate the importance of sociobiology and evolutionary psychology. In the next chapter I will consider some of their greatest successes in explaining human sexuality. But they are looking at only part of the picture. Certainly, much of our behaviour has been selected because it effectively propagates the genes on which it depends. But behaviour is also driven by memetic selection and has been selected because it effectively propagates the memes on which it depends.

I like to look at it this way. There are two replicators driving the evolution and design of our bodies, brains and behaviour. For some aspects of our lives the genes do most of the driving, and the role of the memes can be safely ignored. In these cases the gene-based approach of sociobiology and evolutionary psychology is a good approximation (though still only an approximation), but in other cases the full picture can be understood only by considering both replicators. I will turn now to some of those other cases.

Memetic drive and Dennett's tower

The two examples I have already given are fundamental to understanding human behaviour. They are the big brain and the evolution of language. I have argued that both depend on memetic driving, and I want now to explain this process further and put it in context. The most important step is to show how and why memetic driving is not just another form of evolution in the service of the genes. Unless this is true, memetics can still be reduced to sociobiology.

Memetic driving works like this. Once imitation arose three new processes could begin. First, memetic selection (that is the survival of some memes at the expense of others). Second, genetic selection for the ability to imitate the new memes (the best imitators of the best imitators have higher reproductive success). Third, genetic selection for mating with the best imitators.

The first step means that new ideas and behaviour begin to spread memetically, from making tools and pots, to dancing, singing and speech. The second step means that the people who are best able to pick up the new memes have more offspring who are also able to pick up the new memes. So everyone tends more and more to imitate the successful memes. The third step means that mate choice is also driven by the memes that are prevalent at the time. The consequence of these processes operating together is that the direction taken by memetic evolution affects the direction taken by the genes. This is memetic driving.

Memetic driving may look at first sight as though it is the same as what is known as the Baldwin effect but it is not and I must explain why.

The Baldwin effect was first described by the psychologist James Baldwin who referred to it as 'a new factor in evolution' (Baldwin 1896). It explains how intelligent behaviour, imitation, and learning can all affect selection pressure on the genes. As we have seen, there is no Lamarckian 'inheritance of acquired characteristics' in the sense of passing the results of learning on to the next generation through the genes. Behaviour, however, does have effects on natural selection.

Imagine, for example, a salamander-like creature that eats flies. Individuals that can reach highest get the most flies. Now imagine that one of them begins to jump. It gets more flies and soon any of its mates that cannot jump at all begin to starve. Therefore, genes for good jumping or strong back legs spread in the gene pool and soon all the creatures are more like frogs. Jumping improves and the selection pressure now favours even higher leaps. Behaviour has, in a perfectly Darwinian fashion, affected selection.

Now imagine that the flies vary in appearance and goodness to eat. Let us suppose that striped flies are inedible while spotty flies are excellent food. Froglings that prefer spotty flies will be at an advantage and so the mechanisms required for preferring spotty flies, such as sensitive spot detectors in the visual system, will spread. However, it might be the case that the pattern on the flies changes faster than frogling evolution can track. In this case it will pay the little froglings to be able to *learn* which flies to eat. Any frogling that cannot learn will be at a disadvantage and so genes for a general ability to learn will spread. This is the Baldwin effect.

As Baldwin himself puts it – the highest phenomena of intelligence, including consciousness, the lessons of pleasure and pain, maternal instruction and imitation, culminate in the skilful performances of human volition and invention. 'All these instances are associated in the higher organisms, and all of them unite to *keep the creature alive* . . . By this means *those congenital or phylogenetic variations are kept in existence, which lend themselves to intelligent, imitative, adaptive, and mechanical modification during the lifetime of the creatures which have them.* Other congenital variations are not thus kept in existence.' (Baldwin 1896, p. 445, italics in the original.) In more modern terms, genes for learning and imitation will be favoured by natural selection.

Baldwin thus saw that natural selection, without need of the inheritance of acquired characteristics, could account for the evolution of the capacity to learn. The Baldwin effect creates new kinds of creatures that are capable of adapting to change far more quickly than their predecessors. But this is not the only step in this direction. Dennett explains, using his metaphor of the 'Tower of Generate and Test', an imaginary tower in which each floor has creatures that are able to find better and smarter moves, and find them more quickly and efficiently (Dennett 1995).

On the ground floor of Dennett's tower live the 'Darwinian creatures'. These creatures evolve by natural selection and all their behaviour is built in by the genes. Mistakes are very costly (unsuccessful creatures have to die) and slow (new creatures have to be built each time).

On the next floor live the 'Skinnerian creatures', named after B. F. Skinner (1953) who explicitly saw operant conditioning (learning by trial and error) as a kind of Darwinian selection. Skinnerian creatures can learn. So their behaviour is killed off rather than their whole body. If something they do is rewarded they can do it again, and if not they won't. This is much faster because one creature can try many many different behaviours in a lifetime.

On the third floor are the 'Popperian creatures'. They can evolve behaviours even faster because they can imagine the outcomes in their heads and solve problems by thinking about them. They are named after Sir Karl Popper who once explained that this ability to imagine outcomes 'permits our hypotheses to die in our stead' (Dennett 1995, p. 375). Many mammals and birds have reached this third floor.

Finally, on the fourth floor, are the 'Gregorian creatures', named after the British psychologist Richard Gregory (1981) who first pointed out that cultural artefacts not only require intelligence to produce them in the first place but also enhance their owner's intelligence. A person with a pair of scissors can do more than one without; a person with a pen can exhibit more intelligence than one without. In other words, memes are intelligence enhancers. Among such memes are what Dennett calls 'mind tools' and the most important mind tools are words. Equipped with an environment full of tools that other people have made, and with a rich and expressive language, Gregorian creatures can find good moves and evolve new behaviours very much faster than without. As far as we know, we humans are alone on this top floor of the Tower of Generate and Test.

The importance of the Baldwin effect should now be clear. The Baldwin effect is like the escalator that lifts creatures from one floor to the next. If the necessary good trick is stumbled upon by evolution, and if the costs are not too high, then the creatures who have it are more likely to survive. At each step, they change the environment in which they live so that it becomes ever more important to be good at learning, or whatever. And at each step the creatures who are better at learning are, genetically, at an advantage. Although the Baldwin effect is normally discussed just in the context of learning (stepping up to the second floor), it can equally be applied to the evolution of imagination (getting to the third floor) and of imitation (getting to the fourth floor). Indeed, Baldwin himself explicitly includes imitation in his list of capacities that would help a creature to survive.

But all this is in the service of the genes because the behaviours that are learned, and the solutions that are found by imagining problems, are the ones that help with survival and reproduction. The Baldwin effect is essentially a form of Darwinian evolution acting in the interests of the survival and replication of genes. Several theories of coevolution use the Baldwin effect (such as Deacon's, for example), but the theory of gene–

meme coevolution I am proposing here adds the further process of memetic driving.

The point is that everything changes when you arrive at the top floor. And it changes dramatically. This is because imitation creates a second replicator. None of the previous steps has created a second replicator – at least, not one that operates beyond the confines of the individual. For example, Skinnerian learning and Popperian problem-solving can be seen as selective processes, but they are all going on inside one animal's head. The patterns of behaviour and the hypotheses about outcomes that are selected might be seen as replicators, but they are not let loose on the world unless they are copied by imitation and so become memes.

Getting to the fourth floor means letting loose a replicator that spreads from creature to creature, setting its own agenda as it goes. Of course the genes had no foresight. They could not know that selection for imitation would let loose a second replicator, but that is what it has done, and so we enter the phase of gene–meme coevolution. In this kind of coevolution things happen that serve to spread memes whether or not they spread genes – the dog is off its leash and the slaves have rebelled against their former owners. This is what makes the theory different from previous ones and provides alternative predictions. I suggest that the human brain is an example of memes forcing genes to build ever better and better meme-spreading devices. The brain was forced to grow bigger far faster and at much greater cost than would be predicted on the grounds of biological advantage alone, and this is why it stands out so obviously in any comparisons of encephalisation. Theories based only on biological advantage cannot explain why the genes were forced to pay such a high price in terms of energy consumption and the dangers of birth (see Chapter 6). Theories based on memetic advantage can.

You might still argue that in terms of sheer brain size the results are not so very different from the argument based on the Baldwin effect. However, the big difference between the theories should come in terms of the specific direction in which the brain evolves, not just its size. If the memes have replicator power then they should drive the genes to produce a brain that is specifically suitable for replicating them, rather than one that is designed for some specifically genetic purpose. We should be able to derive predictions based on the requirements of the new replicator to see whether the actual human brain fits the bill. This is precisely what I tried to do in the argument for the evolution of language. The brain we have is a brain designed for spreading memes with high fidelity, fecundity, and longevity.

As it turns out, the big brains have been enormously successful for genes too, and humans have colonised almost the whole planet. But need this have been the case? Might not the memes have actually forced the genes into extinction by pushing for ever bigger and bigger brains and extracting too high a price? We cannot know – though it is an odd fact that we are the only surviving hominids. Could the others have gone extinct this way? The ill-fated Neanderthals did, after all, have somewhat larger brains than modern humans. This is wild speculation indeed, but the more serious point is that on this theory we need not take it for granted that big brains, intelligence and all that goes with them are necessarily a good thing for the genes. We could follow Richerson and Boyd (1992, p. 70) in asking, 'What is so *wrong* with culture that it should be really conspicuous in only one species?'

Maybe the genes have only just managed to carry the burden and fight back in time to produce a species that manages a symbiotic relationship between its two replicators. Maybe we should not assume that when an intelligent, meme-using species evolves it necessarily has a long life ahead.

CHAPTER 10

'An orgasm saved my life'

Sex sex sex sex sex sex – sex – sex.

Did you perk up? Did you pay more attention to the start of this chapter than any other? Probably not. I expect you have developed plenty of defences against sex memes. Nevertheless, if you want to sell magazines, television programmes, or books, one obvious strategy is to put the word 'sex' in a prominent position. A count at my local railway station revealed that of 63 magazines on the shelves 13 had the word 'sex' on the cover – and that is ignoring all the ones with erotic photographs, or headlines like 'Naked couples reveal all', 'How would you like to bed this hungry hunk?', and 'An orgasm saved my life'.

According to the American author Richard Brodie (1996), memes that deal with sex, food, and power all press powerful meme 'buttons' because of the importance of these topics in our evolutionary past. And memes that press buttons are successful memes.

Another way of putting it is that genetic evolution has created brains that are especially concerned with sex, food, and power, and the memes we choose reflect those genetic concerns. Apart from the use of the word 'meme', the logic thus far is exactly that of sociobiologists or evolutionary psychologists who assume that the ideas we have, the stories we pass on, the cultural artefacts and skills that we develop, are all ultimately serving the genes. According to sociobiology, culture should reflect genetic concerns, since culture is ultimately *for* the genes.

Yet, in our own society there are many obvious anomalies. Birth rates have fallen dramatically now that many couples think that having two children is quite enough. Some people have decided that they want no children at all, and prefer to devote their lives to their careers or other occupations. Others adopt children who are not biologically related to them and yet bring them up with great care and devotion as though they were their own. Advertisements, films, television, and books encourage us to enjoy sex with multiple partners throughout our adult lives, without any intention of getting pregnant, and teenagers carry condoms around in their pockets. Contraception has not only brought about effective family planning, but also sex for pleasure and sex for spreading memes. Sexually, we do not behave in ways that maximise our genetic legacy. We no longer have sex in order to get the greatest possible number of our genes into the next generation. We do not buy those magazines in order to have babies. We have largely divorced the act, and joy, and marketing of sex, from its reproductive function.

There are two major ways of accounting for this divorce. The first is sociobiology's answer: modern sexual behaviour is still gene-driven and our use of birth control is (from the genes' point of view) a mistake, made possible because the genes could not anticipate how we would use our intelligence. The second is memetics' answer: modern sexual behaviour is meme-driven. Although our basic instincts and desires are still genetically determined, and these desires in turn influence which memes are successful, the memes themselves are now dictating the way we behave.

I am going to explore both these views and consider their strengths and weaknesses. At the risk of gross oversimplification I am going to lump together as 'sociobiology' much of the work on sexual behaviour stemming from biology, sociobiology and evolutionary psychology. In spite of some differences, they all agree that the fundamental driving force for sexual behaviour is natural selection acting on genes. They do not consider a second replicator, and in this respect differ clearly from memetics.

Sex and sociobiology

The essence of the sociobiological view is that the genes have set up a system that has worked historically but is not entirely appropriate for today's situation. The reason is simple enough. Because genes have no foresight, they can never track environmental changes precisely. Natural selection can ensure that organisms are more or less well adapted to the conditions prevailing at the time, and as times change selection pressures change, so that the better adapted organisms survive. This ensures that tracking is quite effective when conditions change slowly – and extinction is always a possibility when tracking fails. But nothing in the evolutionary process can produce precognition. We are in effect, like all other creatures, products of past selection in past environments.

On this sociobiological argument it is not surprising that our behaviour does not always maximise our genetic fitness. Past evolution has given us a brain that is set up to deal with sex, food, and power, and these ideas are prevalent in our society because these factors all contributed to the survival of our genes in the past. We enjoy sex because animals that enjoyed sex in the past passed on their genes. But evolution has also given us intelligence, which has enabled us to work out the function of sex and manipulate things so as to get the pleasure of sex without the costs of child care. The genes could not have foreseen this and so we have no adaptation against contraception – although, if you agree with E. O. Wilson, you might expect the genes eventually to pull in the leash again and somehow prevent us from reducing our birth rates too far. On this argument our present behaviour is simply a mistake.

Life is full of mistakes. Male frogs quite frequently try to mate with other males and in some species even have to make a 'release call' to escape the unwanted – and extremely lengthy – clutch. Homosexuality in many animals, and even in humans, is sometimes interpreted in a similar way, just as a mistake. Birds with elaborate courtship displays can be induced to strut and flutter and sing for stuffed birds or even for just a few appropriately coloured feathers. Male sticklebacks will fight with very simple dummies and even their own reflections. Presumably the mistakes have not been serious enough to warrant the cost of creating more accurate perceptual systems. Courtship rituals have proved a good way to get a mate even if you occasionally end up dancing and singing for a pile of feathers.

Eating inedible things is another common mistake that is not worth the cost of eliminating completely. Most species survive with very crude systems for distinguishing food from non-food. Chicks will peck at anything of roughly the right size on the ground in front of them, and frogs' tongues will dart out at any small object that moves in the right way. They generally get by perfectly well unless some devious experimenter comes along to trick them. We modern humans have far better visual systems and rarely make such crude mistakes, but we make equally dangerous ones. Selection during our hunter-gatherer past fitted us well for liking sweet and fatty foods. Fish and chips with dollops of sweet tomato ketchup, followed by apple pie with cream and ice-cream would have been extremely good fuel for a *Homo habilis* or an archaic *Homo sapiens*. So we like those tastes, and we enjoy eating chocolates, and doughnuts, and creamy mashed potatoes with sausages and mustard. This is not healthy for an overfed modern *Homo sapiens*. Such mistakes are common in living things.

On this view, birth control and sex for fun, and many other aspects of modern sexual life, are mistakes which the genes have not eliminated – either because the costs would have been too high or simply because the genes, not having foresight, could not eliminate them. However, even if these are mistakes, sociobiologists would argue that most of our sexual behaviour is not. It is the kind of behaviour that served, in the past, to get our genes into the next generation, and will go on doing so in the future.

We should not underestimate how successful sociobiology has been in addressing this central topic, nor how hard a fight it initially had for acceptance. For many decades the popular view was that human beings were somehow above nature and not subject to the constraints of genes and biology. In sexual behaviour, it was thought, we alone could transcend 'mere' biology and make rational conscious choices about whom we made love to and how. Even though nothing is closer to the propagation of genes than sexual behaviour, the theories of the 1950s and 1960s completely ignored biological facts. They made culture the overriding force but, unlike memetics, had no Darwinian account of how culture could exert such a force. With the advent of sociobiology in the 1970s we could begin to make sense of some of our peculiar sexual proclivities (see e.g. Matt Ridley 1993; Symons 1979).

Love, beauty, and parental investment

Consider mate choice. We may like to think that we chose our lover for reasons that have nothing to do with genes and biology; maybe we just fell in love, maybe we chose rationally because he fitted our notion of a perfect husband, or maybe we chose for aesthetic reasons because – well, because he's gorgeous. The truth appears to be that romance and falling in love are themselves based on our deep-seated tendencies to pick sexual partners in ways that would, in our far past, have enhanced the chances of passing our genes into the next generation.

For a start, just how attractive is your partner? I could make a guess that he or she will be just about as attractive as you are. Why? The logic of what is called 'assortative mating' is very simple. Whether you are male or female you need to get the best mate you possibly can. Inasmuch as beauty is relevant to what is 'best' you will go for the most beautiful partner you can get. But then so will everyone else. The result should be, on the average, that people pair up with partners that more or less match them in attractiveness, and this is just what has been found in experiments.

But what is beauty? What makes a man or a woman attractive? The simple answer seems to be that men find women attractive when they have all the signs of being young and fertile, while women are more interested in the status of a potential lover than in his physical appearance. This turns out to have a good biological basis – if a rather complex one.

The basic difference between being male and being female is that females produce the eggs and males the sperm – indeed this is the usual definition of the sexes in widely diverse species. Eggs are large and contain food for the growing embryo, so they are expensive to make, while sperm are tiny and relatively cheap. Eggs are therefore made in smaller quantities and need to be guarded, while sperm can more readily be squandered. In addition, many females also provide a great deal of parental care beyond the provision of an egg, and it is parental care that really makes the difference when it comes to choosing a mate.

The logic of parental investment was initially worked out by the biologist Robert Trivers (1972), following Fisher (1930) who called it 'parental expenditure'. Trivers showed how the sexual behaviour of many different species could be explained by considering how much resources each sex puts into bringing up the offspring. This new understanding was then applied to human behaviour by early sociobiologists. Humans are an interesting case, complicated by the fact that our babies require intensive care for many years and are still unable to fend entirely for themselves for years after weaning. Male parental investment is high compared with other mammals, with males providing food and protection for the family. However, male investment is still much lower than female investment in both traditional and industrialised societies. In existing hunter-gatherer societies women have been shown to provide far more of the nutritionally valuable food for their children than men do and to spend many more hours per day working. Even in our supposedly emancipated Western societies some estimates suggest that women work on average twice as many hours per day as men – that is, including paid jobs, housework, and child care. This disparity in parental investment can explain a lot about human sex.

A human female can produce a maximum of roughly one baby a year in her fertile years, amounting to perhaps twenty or twenty-five in a lifetime. The highest ever recorded is supposedly sixty-nine babies, mostly triplets, born to a nineteenth-century Moscow woman. However, human babies require enormous amounts of care, and in traditional hunter-gatherer societies women probably produced one about every three or four years, spacing the children, as hunter-gatherers do today, by sexual abstinence, long lactation and sometimes infanticide. The simple fact is that a woman cannot increase the number of children she can successfully raise by mating more frequently or with more men.

By contrast a man can, potentially, produce a huge number of offspring. The more women he can impregnate the more babies he will have, and he can more or less rely on the mother to give them some care. Even if some do not survive he will only have invested a few sperm and a brief (and probably pleasurable) effort in producing them. From this simple unfairness follows a world of sexual difference.

For men, the most obvious strategy for passing on the most genes is simply to mate as often as you can with whomever you can. A common and effective method is to have one long-term partner whom you protect as well as you can from other men and whose children you care for, while trying to impregnate as many other women as you can, ideally without getting caught.

Women will pass on the most genes if they can raise a few high-quality children with sufficient resources and care to bring them up. This might mean: (a) mating with a high quality male (i.e. one with good genes), and (b) finding a male who will provide a lot of parental care. These may not always be the same man.

One consequence of this disparity is that females need to be far more choosy about whom they mate with. They do not want to get lumbered with being pregnant by a lazy, ugly, feeble or unhealthy man who will provide poor genes and no care and support for them or their infants. This could explain why women proverbially hold back from sex, and more often have to be persuaded or bought presents. Men do not need to be so choosy. If they can impregnate almost any female it will be worth the small effort because they will not be left holding the baby. This can explain why men are normally much more keen to find sex than women are, and a woman who wants sex can usually find it without too much trouble – and indeed can get paid for it.

Many people seem to hate the idea that their own sexuality comes down to such a crude calculation, but the evidence is piling up against the early anthropologists' idea that sexual behaviour is completely different in each culture. More thorough research now shows that men and women conform to what one would expect based on the principle of higher female parental investment. Men are keener to have sex and are especially turned on by the idea (or the actuality) of sex with many different partners, while women are more choosy and prefer one reliable partner. Prostitution throughout the centuries has been almost exclusively a service rendered by females and paid for by males.

But what about female beauty? Although a man may not lose much by mating with almost any woman, his genes will fare best if he can impregnate a young, healthy and fertile woman. Evolutionary psychologist David Buss found that in every one of thirty-seven cultures, males preferred younger mates and females preferred older ones (Buss 1994). This desire for youth and fertility may explain one of sociobiology's oft-derided findings that men prefer women with a low waist-to-hip ratio

(Singh 1993). Cultures vary in the preferred fatness or thinness of women – with our present obsession for thinness being rather exceptional – but there is apparently a consistent preference for women with narrow waists and broad hips. The reasons for this are still disputed but broad hips appear to suggest a wide birth canal for safely delivering a big-headed baby (of course they may only suggest it and the fat be a deception). A small waist suggests that the woman is not already pregnant, and the last thing a male should want is sex with a pregnant female who may trap him into caring for another man's baby.

Large, clear eyes, smooth skin, fair hair, and symmetrical features are good indicators of youth and health – fairness because in fair-skinned peoples hair colour darkens with age, and symmetry because the effects of disease are often to create asymmetrical blemishes. A long genetic history has created men who respond with sexual arousal to the signs of a young and fertile woman (Matt Ridley 1993).

Meanwhile, the woman need be less concerned with beauty and physical appearance. Her need is for a high-status male who will prove to be a good protector and provider. This fits with the frequent (if depressing) observation that rich and powerful men pair up with young beautiful women. It also fits with the results of surveys showing that men consistently rate physical appearance as important in a partner while women are more impressed by signs of wealth and status. Indeed, looks are critical in women but not men. In all the cultures studied by Buss, the men placed more value on a woman's looks and women placed more value on a man's financial prospects.

But is this the whole story? What exactly is it about men that women find attractive? According to evolutionary psychology, the genes determining female mate choice should be those that would have been selected for in a hunting and gathering life. In that way of life there are few possessions because people are always on the move, but providing regular supplies of meat and useful tools would presumably help in providing for children. Status might be earned by prowess at hunting, or fighting or protecting the group from enemies, or perhaps by impressive clothing or decoration. Can the genes for choosing these qualities in a man really lead us nowadays to choose men with big bank balances, fast cars, high-profile jobs, and beautiful houses? Possibly, although as we shall see, memetics takes a different view.

Another important biological fact is that a woman can be certain that her baby is her own, and may have a good idea of who the father is. Men cannot (or could not until the advent of genetic fingerprinting). This difference is especially acute in humans because women, rarely among primates, have concealed ovulation – neither they nor their partners know at what times of the month they are fertile. A man cannot guard a woman all the time; so she may be able to trick him into caring for another man's child. Indeed, this may account for the evolution of concealed ovulation (R. R. Baker 1996).

There are many ways for a man to increase the probability that he is the father of the child he is feeding or protecting. Marriage is one, and it is reinforced by men insisting on premarital chastity and marital monogamy. Some of human beings' nastiest practices (at least from the female point of view) may also serve to increase paternal certainty, such as the mutilation of female genitals, chastity belts, punishment of women (but not men) for adultery, and various methods of locking women away from the world. I suppose I fell foul of the same unfairness in a minor way in the early 1970s. During my first term at Oxford I was unfortunate enough to get caught with a man in my room at eight in the morning. The man concerned was fined two shillings and sixpence (about twelve pence in today's money, and not much even then) and told to be more careful by his 'moral tutor'. My parents were summoned to college and I was sent away from the University for the rest of the term.

If parental certainty is so important, jealousy should serve different functions in men and women. The evolutionary psychologists Martin Daly and Margo Wilson argued that if what men fear most is being cuckolded they should be especially jealous of their partner's sexual infidelity, whereas if what women fear most is desertion they should be most jealous of their partners spending time and money on a rival. Many studies show that this is exactly so (Wright 1994). David Buss even wired people up with electrodes and asked them to imagine their partners having sex with someone else or forming a deep emotional attachment to someone else. For men it was the sex that caused all the physiological signs of distress; for women it was the emotional infidelity (Buss 1994).

Finally, there is a last wicked twist to the argument. Women certainly want to get as much male investment as possible, but they may not be able to find both good genes and a good provider in the same man. Indeed, a man with good genes – tall, strong, and intelligent, for example – may find it so easy to get sex that he need not bother with putting effort into child care. This is apparently true in zebra finches and swallows where the more attractive males have been shown to work less hard in bringing up the young, leaving the females to work harder. On the 'best of both worlds' theory a woman's best bet may be to capture a nice, though unattractive man, who will rear her children, and then go and get better genes from elsewhere. As Matt Ridley (1993, p. 216) puts it 'marry a nice guy but have an affair with your boss'.

We can probably all think of examples, but can such behaviour be biologically effective in modern humans? Evidence that it can comes from controversial research by British biologists Robin Baker and Mark Bellis (1994; Baker 1996). In a survey of nearly four thousand British women they found that women who were having extramarital affairs tended to have sex with their lovers more often when they were ovulating – and this was not true for sex with their husbands. In addition, they had more sperm-retaining orgasms (i.e. orgasms between one minute before and forty-five minutes after the man's) with their lovers than with their husbands. In other words, if they were not using contraception they might still be more likely to get pregnant by the lover even though they had sex with him less often.

These are just some of the ways in which modern sociobiology and evolutionary psychology are coming to understand human sexual behaviour and mate choice. Some of the details may prove to be wrong and new theories will come along, but there is no doubt how effective this approach has been. However, there are many things about human sexual life that just do not seem explicable this way and will not, I suggest, succumb to a sociobiological account.

Memes and mates

There are two main ways in which memetic theory differs from a purely sociobiological account of sex. First, memes have been around for at least 2.5 million years; coevolving with genes and influencing sexual behaviour and mate choice. Second, memes are now well off the leash and during the last century sexual memes have influenced our lives in ways that have little or nothing to do with genes.

Let us begin with mate choice. The main difference between the two theories is this. According to sociobiology our choice of mates, and whom we find attractive, should ultimately come back to the question of genetic advantage. Our modern life may complicate things, but essentially we should choose to mate with people who would, in the environment of our evolutionary past, have helped to increase our genetic legacy.

According to (my version of) memetics, mate choice is influenced not only by genetic advantage but also by memetic advantage. One of my key assumptions has been that, once memes arose in our far past, natural selection would have begun to favour people who chose to mate with the best imitators or the best users and spreaders of memes. This was part of my argument for the memetic driving of genes for bigger brains and language, but it also leads naturally to some conclusions about mate choice. As memetic competition took off in our far past, so the direction the memes took would have affected mate choice. People would have tended to mate with the best meme-spreaders, but what constituted the best meme-spreaders depended on what the memes were doing at the time. It is in this sense that the memes began to call the shots.

Let us consider some examples. In an early hunter-gatherer society a man who was especially good at imitation would have been able to copy the latest hunting skills or stone tool technology and hence would have gained a biological advantage. And a woman who mated with him would be more likely to have children who shared that imitation ability and that advantage. So how would she choose the right man? I suggest she would have to look for signs, not just of having the good tools because they might change, but of being a good imitator in general. This is the critical point – in a world with memes, signs of being a good imitator change as fast as the memes change. Genes for choosing men who could make and use the old stone tools might once have had an advantage but as more memes arose and spread they would not. Instead, genes for choosing men with the general ability to imitate, or even to innovate, would fare better. In a hunter-gatherer society such signs might include making the best tools, singing the best songs, wearing the most stylish clothes or body paint, or appearing to have magical or healing powers. The direction which memetic evolution took would have influenced the genes.

If this argument is right we would expect the legacy of memetic driving to be visible in our mate choice today – that is, we will still mate with the best imitators (and to some extent the best imitators of the kinds of memes that have been around in our past). In a modern city, clothes fashions might still be one sign, but others would include musical preferences, religious and political views, and educational qualifications. More important, though, would be the general ability to spread memes – to be the fashion setter as well as the best follower. This suggests that desirable mates should be those whose lives allow them to spread the most memes, such as writers, artists, journalists, broadcasters, film stars, and musicians.

There is no doubt that some of these occupations give you a good chance of being mobbed by admirers and of having sex with almost whomever you like. Jimi Hendrix apparently fathered numerous children in four countries before he died at the age of twenty-seven. H. G. Wells, although notoriously ugly and with a squeaky voice, reputedly specialised in seducing several women in one night. Charlie Chaplin was short and far from good looking but a great sexual success story – as, apparently, were Balzac, Rubens, Picasso, and Leonardo da Vinci. The biologist Geoffrey Miller argues that artistic ability and creativity have been sexually selected as a display to attract women (Miller 1998; Mestel 1995), but he does not explain why sexual selection should have picked on these features. Memetics provides a reason – that creativity and artistic output are ways of copying, using and spreading memes, and hence are signs of being a good imitator. I would predict that if these things could be teased out, women would, other things being equal, prefer a good meme-spreader to just a rich man.

Note that I have couched this argument in terms of female mate choice. There is some sense to this because, as previously discussed, females need to be more choosy over mates than males do, and, in general, sexual selection is driven by female choice – as in the examples of peacocks' tails and other fancy plumage. However, this imbalance is not necessary for the argument I am pursuing here, and we may find that men too have tended to mate with women who are good imitators. Also, in today's technologically advanced societies, women can spread memes as well as men can. So we may expect many more changes in sexual behaviour and mate choice as women increasingly take control over the spread of memes.

My suggestion that we should mate with the best imitators is central to the theory of meme–gene coevolution and memetic driving; so it is an obvious one for testing. The predictions are quite straightforward: that people should choose mates according to how good they are at copying, using, and spreading memes. Experiments might be designed to hold genetic factors constant, and manipulate memetic factors, while measuring perceived attractiveness. More subtly, we might explore the interactions. I would expect that an ugly and impoverished man might still be perceived as attractive if he were a great meme-spreader – but just how much ugliness could one get away with? Even in today's meme-rich society, women very rarely choose men who are shorter than they are. Obviously, there is a limit to how far memes can overthrow genetic considerations, and this provides a fascinating area for research.

Memes are now spreading farther and faster than ever before and this has powerful effects on everything in our lives, including sex. The second way in which memetic theory differs from sociobiology is in the way it accounts for sex in the modern world. It is time to return to those sexy magazines and the quandaries of celibacy, adoption, and birth control.

CHAPTER 11

Sex in the modern world

It is time to come into the twentieth century. I have spent much of this book explaining how memes came about in human evolution and how they might have pressed the genes into producing a creature with an exceptionally big brain and the capacity for language. For most of this long evolutionary time our hominid ancestors had few memes to play with. They lived in relatively simple societies and there was little communication between distant groups. Things are not like that now. Not only are there far more memes in circulation, but the way they are passed on has changed.

Many memes are passed from parent to child. Parents teach their children many of the rules of their own society; how to hold chopsticks or a knife and fork, what to wear on which occasion, how to say please, thank you, no thank you, and countless other useful things. Children get their first language from their parents and usually their religion too. Cavalli-Sforza and Feldman (1981) call this vertical transmission, as opposed to horizontal transmission (between peers), or oblique transmission (e.g. from uncle to niece or older to younger cousin). The mode of transmission is important because it affects the relationship between memes and genes.

When memes are transmitted vertically they are transmitted alongside the genes. In general, this means that what benefits one also benefits the other. So, for example, if a mother teaches her child how to find food, how to avoid danger, how to dress up to look attractive, and so on, then she is not only helping her child survive but also helping the propagation of her own genes and her own memes. Indeed, if all transmission is vertical there can be little conflict between memes and genes (and no need for memetics). The sociobiologist's leash is very tight indeed and all the memes that are created should be expected to be, at least in principle, helping the genes. In fact, they may deviate from this ideal in all sorts of ways, and fail to track the genes perfectly, but the principle is clear. When you pass on your ideas to your child it is in your own genetic interest to pass on ideas that benefit that child's reproductive success. And from the meme's point of view, its survival also depends on your reproductive success.

This suggests that coevolution and memetic driving cannot occur with only vertical transmission of memes. So we should note at this point that all the examples I have given of memetic driving involve at least some element of horizontal transmission. I suggested, for example, that people imitate the best imitators. This entails horizontal or oblique transmission, as does the creation of language, since you cannot easily have a language community in which people speak only to their parents and children.

Where memes are transmitted horizontally they can travel quite independently of the genes. An idea may be passed from one person to another, and then another, within one generation. Also memes may spread when they are useful, neutral or even positively harmful; such as an untrue explanation, an addictive habit, or a piece of malicious gossip. Only when horizontal transmission becomes common can the memes really be said to be independent of the genes.

Modern industrialised life is a world of horizontal transmission. We still learn our mother-tongue from our mothers, and many of our habits and ideals too. People are still overwhelmingly more likely to follow the religion of their parents than any other religion, and even to vote the same way as their parents did. However, our parents have less and less sway as we get older and we now go on learning more or less throughout our lives. Our main sources of information are sources that did not exist in our long evolutionary past: schools, radio, television, newspapers, books and magazines, and lots and lots of friends and acquaintances widely spread around the city, the country, and even the world.

The more ways there are for memes to spread, and the faster they can go, the less they are constrained by the needs of the genes. What determines the success of a meme in a traditional hunter-gatherer society, or even a simple farming society, is quite different from what determines its success in a modern industrialised society. In the former, life changes slowly, transmission is largely vertical, and a meme is most likely to succeed if it benefits (or at least appears to benefit) the health, longevity, and reproductive success of its carrier. In the latter, a meme is most likely to succeed if it can get quickly and efficiently from host to host, and never mind how well each host does in terms of either its own survival or its reproductive success – as long as there are more hosts around to infect. We now live in this latter kind of society and the memes have utterly changed – and continue to change – the way we live.

We can now return to the subject of sex and sex memes. For the sake of simplicity I am going to divide societies into just two types, realising that there are lots of gradations in between and few are purely one or the other. These are societies in which memes are largely transmitted vertically and thus track the genes, and those in which they are transmitted horizontally and do not.

First, let us consider vertical transmission. There are many memes that ride on the back of biologically determined behaviours. They include all the memes that exploit biologically determined tendencies for mate choice and other aspects of sexual behaviour. From the examples given in the previous chapter we can easily guess at a lot of them: pictures of beautiful women with slim waists, long fair hair, bright eyes and symmetrical faces; films and videos of other people having sex, or of stories about other things with plenty of sex thrown in. Because people want to see these images, money can be made from them. Stories about jealous husbands and abandoned women will sell, as will love stories about pretty young nurses and clever, successful doctors (if you think these are a thing of the past go and look in the romance section of your local bookshop!).

Memes concerned with marriage are another obvious example. From fluffy white dresses and bunches of flowers, to defloration ceremonies and horrible punishments for adultery, we can understand many of the memes that surround marriage as being grounded in biological advantage. The American memeticist Aaron Lynch (1996) has provided many examples of marriage traditions that track biological advantage, including gender roles and patrilineal inheritance. The mechanism here is simple. People who practise a certain kind of marriage system will produce more children than those practising another, and so will pass on that system to their own, more numerous, children – so spreading that practice more effectively.

Moreover, the system that works best may vary with the environment. Socio-ecologists have provided many examples of unusual marriage arrangements, and of the varieties of bride prices and dowries, that actually seem to track the environment and enhance genetic fitness for the people who practise them. Polygyny (one man having several wives) is a common system, as is monogamy. But in extreme environments other systems can prevail. For example, the marginal, cold, and infertile valleys of the high Himalayas are one of the very few places in the world where fraternal polyandry occurs, that is, one woman marries two or more brothers who inherit the family land. Many men and women remain celibate; women usually helping on the estate, and unmarried men becoming monks. The British socio-ecologist John Crook (1989) has studied these people in detail and argued that their system does, in fact, maximise their genetic fitness. Grandmothers with polyandrous daughters were found to have more surviving offspring than those with monogamous daughters (Crook 1995).

Whether you look at this from a sociobiological perspective, or a memetic one, the outcome is similar. The successful practices (or successful memes) are those that provide the greatest genetic advantage in the given environment.

The same is true for some widespread sexual taboos. Masturbation has been seen as dirty, disgusting, revolting, and as sapping your 'vital energy'. Generations of boys have been brought up to believe that 'playing with themselves' will make them go blind, or give them warts, or make hair grow on the palms of their hands. Given that young men have a strong desire for sex, dissuading them from masturbation is likely to increase the amount of vaginal sex they will have, thereby increasing the number of their offspring to whom they can pass on the taboo (Lynch 1996). Lynch suggests a similar explanation for the success of the circumcision meme, because circumcision makes masturbation more difficult, but not vaginal sex.

Interestingly, there are few, if any, taboos against female masturbation. Recent research shows that, though women masturbate less often than men, many masturbate once a week or more throughout most of their adult life (R. R. Baker 1996). The lack of a taboo makes sense because generally women cannot increase the number of their offspring by having more sex, so from this point of view it does not matter whether they masturbate or not.

The taboo against homosexuality follows the same logic. Most homosexuals are at least partly bisexual and can, with strongly wielded taboos, be persuaded to marry and have children, to whom they will pass on the taboo. Similarly, taboos against any kind of sexual practice that does not involve insemination can spread, including those against using birth control. Taboos against adultery work rather differently. Brodie (1996) suggests that it is in every man's genetic interest to persuade other men *not* to commit adultery while doing so themselves. Thus both the anti-adultery memes and hypocrisy spread together.

Finally, there are many religions that make use of sex to spread themselves. A religion that promotes large families will, assuming vertical memetic transmission, produce more babies to grow up in that religion than one that promotes small families. Religious memes therefore become an important manipulator of genetic success. Catholicism's taboo against birth control has been extremely effective in filling the world with millions of Catholics who bring up their children to believe that condoms and the pill are evil, and that God wants them to have as many children as possible.

Note that I said 'assuming vertical memetic transmission'. All the above arguments depend on parents passing on their memes to their children, because only in this case does the number of children you have determine the success of your memes. Vertical transmission was probably a major route of memetic replication throughout most of our evolutionary history. Early humans probably lived in groups of about one to two hundred at most. They may have communicated with many of the group, but they would have been unlikely to communicate much more widely than that. As far as we can tell, cultural traditions changed very slowly for thousands of years and so the memes that parents passed on to their children would have continued to be the prevalent ones throughout the children's lifetime. In this situation, the successful memes would, to a large extent, be the ones that were also of biological advantage.

In examples like these the sociobiological and memetic explanations barely differ. They do not make different predictions. There is no particular advantage to the memetic viewpoint, and we might as well stick with sociobiology.

However, transmission is no longer largely vertical. So what happens to sex when memes are generally spread horizontally? The simple answer is that biological advantage becomes less and less relevant. Let us take the first type of sex meme that I mentioned: the pictures of sexy women and the heart-rending love stories. These are not affected because they depend on biologically inbuilt tendencies that will not quickly go away. Even though we now spread most of our memes horizontally, we still have much the same brains as people did five hundred years ago or even five thousand years ago. We just do like tall, dark, strong hunks, and slim, bright-eyed females. We just are turned on by watching sex or thinking about our ideal lover while masturbating.

The same is not true of social institutions like marriage practices. Nowadays, what determines the memetic success of a marriage practice is not how many children it produces. Horizontal transmission is now so fast that it outstrips vertical transmission and people can choose what kind of marriage system to adopt from any they happen to have come across, including none at all. The number of children produced by their parents' marriage system is now irrelevant. Monogamous marriage has survived a long time and is still prevalent even in technologically advanced societies. But it is clearly under pressure, with divorce rates reaching nearly 50 per cent in many countries, and some young people rejecting the 'ideal' of marriage altogether.

I mentioned the rare practice of fraternal polyandry which increases genetic success in some parts of the Himalayas. With increasing access to city lifestyles and more horizontally transmitted memes we might expect such a system to break down, and indeed this is exactly what is happening. As remote Himalayan villages come into contact with the rest of the world, young men are increasingly choosing not to share a wife with their brother but to opt for city life instead (Crook 1989).

The taboos also are no longer as effective as they once were. We may imagine a 'masturbation taboo' meme competing with a 'masturbation is fun' meme. The question of how many children the carriers of such memes produce is now completely irrelevant. People will pick up their memes from films, radio, books and television long before they have even produced any children, let alone persuaded their children to copy their own habits. So we should expect the power of all these sexual taboos to be reduced as horizontal transmission increases – as indeed appears to be the case.

The taboo against homosexuality is especially interesting. There is no generally accepted biological explanation of homosexuality and superficially it does not appear to be adaptive. Nevertheless, evidence is accumulating that there is an inherited predisposition for homosexuality. Assuming this is the case, the taboos of the past would, paradoxically, have favoured the survival of these genes by forcing the people who carried them, against their wishes, to marry and have children.

This suggests an interesting prediction for the future. As horizontal transmission increases the taboo should lose its power and so can be expected to disappear, as indeed it is doing in many societies. Homosexuals are then free to have sex with other homosexuals, to have long-term relationships with their own sex, and not to have children at all. The short-term effect is much more overt homosexual behaviour and acceptance of that behaviour by everyone, but the long-term effect may be fewer genes for homosexuality.

This analysis suggests that ancient sexual taboos should disappear, not as a function of wealth or industrialisation *per se*, but with increasing horizontal transmission. Thus, we would expect cultures with the least horizontal transmission to have the strongest taboos and vice versa. There are many indirect measures of horizontal transmission such as literacy rates, or the availability of telephones, radios and computers. More direct measures would be estimates of the average social group size, or the number of contacts that people make with others outside their immediate family. I would expect negative correlations between all these measures and the prevalence of sexual taboos. In this case, memetics provides predictions that do not make obvious sense within any other framework.

Celibacy

We can now return to those aspects of modern life that I suggested provided a special challenge to sociobiology; celibacy, birth control, and adoption.

Why would anyone voluntarily remain celibate and forego all the pleasures of sex? Unless they are constructed entirely differently from the rest of us, they will presumably have to fight hard against the natural desire to have loving physical relationships and to relieve the occasional, or even persistent and desperate need for sex. Celibate people cannot, by definition, pass on their genes. So why do they do it?

Genetic explanations are not impossible. Celibate men or women might, under some circumstances, better promote the survival of their genes by caring for siblings or nieces and nephews. This is known to happen in some territorial birds. For example, when territories are scarce young unmated males help at the nests of their siblings. They may get a territory of their own in future seasons, but for now helping out their nieces and nephews may genetically be the best bet. Certainly, among humans the loving maiden aunt and generous bachelor uncle are well known, and nepotism is common enough to warrant having its own name. Also, we have already considered one marriage system in which many people remain celibate but their genes still do better because of the impoverished environment.

So genes and environment might account for some kinds of celibacy but what about the celibate priest in a wealthy society? He cannot have inherited the celibate lifestyle genetically. He is unlikely to spend his time tending his brothers' children and grandchildren, and his absence from the family is unlikely to benefit them by leaving more food for them to eat. If he is truly celibate (and, of course, many are not) his genes will die with him. Religious celibacy is a dead end for genes.

Richard Dawkins gave the first memetic explanation of celibacy in *The Selfish Gene* (1976). Suppose, he said, that the success of a meme depends on how much time and effort people put into propagating it. From the meme's point of view any time spent doing anything else is simply time wasted. Marriage, having children and bringing them up, even sexual activity itself, is a great waste of time for memes. Suppose, he went on, that marriage weakened the power of a priest to influence his flock because his wife and children occupied a large proportion of his time and attention. Then it follows that the meme for celibacy could have greater survival value than the meme for marriage. A religion like

Roman Catholicism, which insists on celibacy in its priesthood, would find itself with actively meme-spreading priests, plenty of converts and an ongoing supply of new recruits to celibacy. The agony of abstinence may even goad these priests into ever more fervent attempts to serve their religion as a way of diverting their own attention from wicked thoughts of sex.

This is a particularly interesting kind of meme–gene conflict reminiscent of the gene–gene conflicts between a host and a parasite. I already gave one example in the conflict over the thickness of snail shells. Some parasites actually castrate their hosts (usually chemically rather than physically) as a way of diverting the hosts' energies into replicating them rather than host genes. Religious celibacy is a way for memes to divert their hosts' energies into replicating religious memes instead of host genes (Ball 1984).

If this explanation is going to be really useful it should be able to predict the conditions under which celibacy will and will not evolve, and I shall return to this when considering religions in more detail. For now the point is clear enough. Memetics suggests that some behaviours will spread just because they are good for the memes. You could look at it this way – each person only has a finite amount of time, effort and money. Their memes and genes therefore have to compete for control of these resources. In the truly celibate priest the memes have won hands down. But they have not done badly in even a lapsed celibate priest. As we know from many recent scandals, quite a few priests do have affairs and become fathers. But, of course, they have to keep this secret. They do not usually give up their religious life and so they cannot afford to spend any time and effort, or much money, on those offspring. They must rely on the mother providing all the care. If she does, the sinful man's memes and his genes will both have done well.

Birth control

Birth control succumbs to exactly the same argument – and with dramatic consequences for the future of both memes and genes.

Let us suppose that women who have many children are far too busy to have much social life, and spend most of their time with their partners and family. The few other people they do see are likely to be other mothers with young children who already share at least some of their child-rearing memes. The more children they have the more years they will spend this way. They will, therefore, have little time for spreading their own memes, including the ones concerned with family values and the pleasures of having lots of children.

On the other hand, women who have only one or two children, or none at all, are far more likely to have jobs outside the home, to have an exciting social life, to use e-mail, to write books and papers and articles, to become politicians or broadcasters, or do any number of other things that will spread their memes, including the memes for birth control and the pleasures of a small family. These are the women whose pictures appear in the media, whose success inspires others, and who provide role models for other women to copy.

There is a battle going on here – a battle between memes and genes to take control over the machinery of replication – in this case a woman's body and mind. Any one person has only so much time and energy in their lifetime. They can divide it as they choose but they cannot have lots of children *and* devote maximal time and effort to spreading memes. This particular battle is played out largely in the lives of women and is becoming ever more important as women take a more prominent role in modern meme-driven society. My argument is simply this – the women who devote more time to memes and less to genes are the more visible ones, and therefore the ones most likely to be copied. In the process, they are effectively encouraging more women to desert gene-spreading in favour of meme-spreading.

This simple bias ensures that memes for birth control will spread even though they are disastrous for the genes of the people carrying them. These memes include not only ideas about small families and the benefits of birth control, but the pills, condoms, and caps that actually do the work; all the ideas in our society concerning sex for fun; the films, books, and television programmes that promote them; and the programmes of sex education that help our children to cope with sex in a permissive society without getting pregnant or catching AIDS. If this theory is right, birth rates are unlikely to rise again because this simple bias will keep them down.

Is this theory right? It makes a number of assumptions that could be challenged. One crucial assumption is that women with fewer children copy more memes. This seems to be true in a world in which middle class women with more money and more access to information have fewer children, but it could easily be tested by measuring, for example, the number of social contacts they have, the time they spend talking to others, the amount they read, their output of written or broadcast material, how many of them use e-mail, or own fax machines. The theory can work only if memetic output correlates negatively with the number of children a woman has.

A second assumption is that women are more likely to imitate the women they see in the media who have (or appear to have) few children, than those of their friends who have many. Research in social psychology, marketing, and advertising shows that people are more often persuaded by others who are perceived as powerful or famous. Family size is probably no exception, so if successful women have few children then others will copy their example. If both these assumptions are true then it follows that, in a climate of horizontal transmission, birth control will spread and families will get smaller.

Predictions can also be made. For example, the size of families should depend on the ease with which memes can spread horizontally in a given society. Other theories might predict that the main forces for lowering birth rate (Chinese-type coercion aside) would be economic necessity, availability of birth-control technology, the value of children as agricultural workers, or the decline of religions. Meme theory suggests that factors such as how many people a mother typically communicates with, or how much access she has to printed and broadcast material, should be more important. And note that it is mothers who count. Memetic theory easily explains why the education of women is so important in changing family size.

Education aside, this all leads to the paradoxical thought that the more sex magazines, e-mail sex sites, and sex shops are available, the lower birth rates are likely to be. The sale of sex in modern societies is not about spreading genes. Sex has been taken over by the memes.

Let us consider an example. Imagine a couple who both have rewarding and demanding careers. Let us suppose that she is editor of a magazine and he is a management consultant. They have a large house but it is a workplace as well as a home. They have computers, fax machines, phones, and desks piled with work, and they work long hours. She goes to the magazine's office, but often works at home, editing contributions, dealing with problems and writing her own articles. When they are not working they go out with friends to get a welcome break from it all.

The time comes to decide about children. The woman is in her thirties; she has always faintly wanted children, but how will she manage? She sees her friends juggling family and careers, she sees the time that babies take up, and the sleep they deprive you of, the problems with nannies, the money they cost. She thinks about her work: they are about to take over another magazine. Will she get the job of editing both? If she takes time off, will she lose it? He thinks about his clients. Will children be in the way? Will he need a separate office? Will his competitors overtake him if he cannot keep working evenings and weekends? What if he has to take children to school or do his share of nappy-changing and feeding? They decide, on balance, not to have children.

What has happened here? You could say that these two people have rationally made the choice to devote their energies to work rather than having children. And in one sense you would be right. But another way of looking at it is from the meme's eye view. From this perspective the memes have done rather well. They have, as it were, persuaded the couple to devote their energies to memes rather than genes. They did not do this by conscious design or foresight, but simply because they are replicators. From this perspective, the couple's thoughts, emotions, desire for success, and willingness to work hard, are all aspects of the replicating machinery that is, or is not, devoted to spreading the memes – as are the printing presses that reproduce the magazines and the factories that build the computers. The buyers of the magazines and the users of the management advice are all part of the environment in which all these memes thrive, and these memes use us for their own propagation.

There are many people like this. As our environment becomes more and more rich in memes and meme-copying devices, we may expect more and more people to become infected with memes that drive them to spend their lives propagating those memes. That is what memes do.

The overworked scientist is frantic to read all those latest research reports. The exhausted doctor cannot keep up with the latest change in health care advice and works longer and longer hours. The advertising executive has a mountain of new ideas to deal with. The check-out worker at the supermarket has to learn the latest technology or lose the job. With the advent of the Internet more and more people are getting connected and there is scope for them to spend inordinate amounts of time playing with the new memes. The computer nerd is more in the thrall of the memes he plays with than of the genes he is carrying.

The natural end point of all this might appear to be a childless society, but the genes have given us a powerful desire to have and care for children. I would guess that birth rates in modern meme-driven societies will stabilise at some level that balances the genetically created desire for children against the memetically created desire to spread memes more than genes.

Adoption

Finally, there is the question of adoption. Sociobiologists can reasonably argue that childless couples are driven by their genetically created desires to want to have and nurture children, and that these desires outweigh the obvious point that an adopted child will not pass on their genes. In other words, from the gene's point of view adoption is just a mistake. However, it is an extremely expensive mistake. It means devoting vast amounts of time and money for no genetic reward at all. It is just the kind of mistake that is made by the bird who is tricked into bringing up a baby cuckoo, or a man who is 'cuckolded' – and we have already seen how far biological evolution has gone in devising strategies to avoid this happening – and what pressures men have put on women to ensure their own paternity. Genetically, infertile people would do better to help their siblings and their siblings' children. Some do just that, but the long queues of people now waiting to adopt suggest there is something going on here that challenges the sociobiological view.

Looked at from the meme's point of view, the benefits of adoption are obvious. As far as memes are concerned, the time and effort expended on an adopted child are exactly as valuable as that expended on one's own offspring. There are many kinds of meme which parents pass down vertically to their children. The ones that are successful in getting spread this way (and are common in the meme pool) are those that people *want* to pass on. These include not only religious and political views, social mores, and ethical standards (in any case some children reject these entirely), but all the possessions that go with living in a meme-rich society. Memes are ultimately responsible for us having our homes and possessions, our position in society, and our stocks, shares and money. None of these things would exist without a meme-based society and these are the things we work hard for and want to leave to someone we care about when we die.

If we ask someone why they want to adopt a child we should not expect them to say 'to pass on my memes', any more than if we ask someone why they enjoy sex we would expect them to say 'to pass on my genes'. Nevertheless, from the meme's eye view a person's desire to pass on their experience and possessions is an opportunity to be exploited. Thus, we should expect that in species without memes, individuals will do all they can to avoid bringing up non-relatives, but in a species with both memes and genes, some individuals will find themselves wanting a child whether or not it is biologically their own. Adoption, birth control and celibacy may be mistakes for the genes, but they are not so for the memes.

. . . .

The memes can take over sexual behaviour in many other ways too. Sex means intimacy, and intimacy means sharing memes. Many a spy has lured a politician into bed as a ploy for obtaining information. Many a young actress has succumbed to sex on the casting-couch in the hope that she will get on to the wide screen and so be seen and possibly emulated by millions. Power is a powerful aphrodisiac and today's power is all about spreading memes. Politicians are renowned for using sex as a weapon, as a device to gain influence, and as a way of cementing alliances – and these alliances are all about spreading political memes. Sex is a wonderful world for the proliferation, control, and manipulation of memes.

I have contrasted the sociobiological view of sex (it is all for the genes) with a memetic view (it is for memes as well as genes). These two approaches make rather different predictions for the long-term future of any memetic species. If sociobiologists are right (at least those who agree with their founding father, E. O. Wilson) then the genes must ultimately pull in the leash again. If the genes were fundamentally in charge they would find a way to correct the mistake and redress the balance. As time goes on, unless the mistake proved fatal, human beings would change genetically so that they were no longer lured away by magazines, high-powered jobs, or the Internet, and were prepared to concentrate on the proper business of creating more human beings.

There is no room for such a leash on the memetic view. If memes are replicators in their own right then they will spread and spread entirely selfishly. They will also spread faster and faster, and the number of memes will go on increasing. If the genes ever could track the memes there must come a point at which they can no longer do so, and the speed of memetic evolution leaves the genes far behind.

In today's world a very few people still live as hunter-gatherers; many live as farmers or industrial workers in rapidly changing countries; and some live as advanced spreaders of memes in societies with computers, mobile phones, and television. Birth rates are highest in the developing countries and lowest in the technologically advanced ones, so at the moment, memetic pressures favour the genes of people living in undeveloped countries. Since their genes differ very slightly from the genes of people in developed countries, this will have some effect on the overall gene pool. However, for this to have a big effect, the selection pressure would have to remain stable for many generations and, given the rate of cultural change, this now seems unlikely. So what might we expect to happen now?

For most of the past two or three million years memes have evolved slowly. Their main effect on the genes occurred because people tended to mate with good imitators, but beyond that they did not affect sexual behaviour very much. Our sexual behaviour was largely driven by the genes for their own replication, and our sexuality still shows the legacy of this long process. However, in the modern world, the memes have taken over much of our sexual behaviour and put it to work for their propagation instead. The technology of birth control has been an extraordinarily successful set of memes, facilitating the sex industry and diverting people's energies away from lifelong child rearing. However, just like genes, the memes have no foresight. They cannot be expected to foresee that almost anything might happen. They might even, in the process of diverting our energies from the genes, just wipe us out.

Actually, this possibility is remote for the following reason. If birth rates over the whole planet fall then the total population will fall. This will be good news for the rest of the biosphere, but bad news for the memes. At some point, the density of population will be too low to sustain the infrastructure needed for a thriving memetic world, and so memetic driving will slow down, and with it the use of birth control. The genes can then take over and build up the population again until a new memetic infestation takes place. As is seen with the spread of many parasites and diseases, it is rare for them to wipe out their hosts completely, and I would not expect memes to do so either.

In fact, the situation is far more complex and unpredictable than that. Given the gross inequities between societies at the moment it is more likely that birth rates will continue falling in the technologically advanced societies, while the less advanced ones grow in population. Memetic influence may then shift towards the previously undeveloped countries and birth rates start falling there. So we might expect a swing back and forth as memes and genes battle it out to get human beings to spend their lives replicating one or the other. This is what it means to be creatures of two competing replicators.

Finally, the memes are busy devising ways of interfering even more directly with the genes. We are already creating genetically engineered vegetables for food and, memetic pressure groups notwithstanding, will probably create genetically engineered animals that are quick growing, delicious to eat, and do not object to their impoverished or miserable lives.

DNA testing means that paternity can be assured so that women will find it harder to trick their partners into raising other men's children and men will have to pay the cost of having children by casual relationships. Our sexual desires will still follow the dictates of genetic evolution while memetic evolution changes the rules. Genetic engineering is already becoming commonplace, and some of the major inherited diseases may soon be defeated by simply removing the genes that cause them. Cloning of sheep and other large animals is also possible and, combined with the creation of headless and brainless clones, raises the prospect of maintaining genetically identical spare parts for rich people to ensure that they always have a heart or liver ready if they need one. Other predictions for the future of 'reprogenetics' include babies born with the genes from two mothers, the insertion of genes for AIDS resistance into embryos, or the creation of whole suites of synthetic genes to provide designer embryos for people rich enough to afford them (Silver 1998).

Note that I said 'the memes are busy devising'. This translates into the more accurate statement that memes for DNA testing, sequencing the human genome, and genetic engineering are successfully replicating in today's world. Why? Because many memetic factors come together to make them successful. Enough people are sufficiently well educated; there are laboratories full of the necessary equipment; there are clever people around who manage to combine the existing memes and come up with new inventions; there is sufficient wealth to educate and fund those people to do it – and, of course, there is the human desire to have healthy, happy and successful children, and human greed that will always want more and better food, and promises of a better and easier life.

So are we just unalterably selfish creatures, driven by the competing forces of two replicators to live lives of mindless greed? Not at all. Rather surprisingly one of the consequences of memetic evolution is that humans can be *more* altruistic than their genes alone would dictate.

CHAPTER 12

A memetic theory of altruism

Altruism in the service of the genes

Once one of the greatest mysteries for sociobiology – and now probably one of its greatest successes – is the problem of altruism.

Altruism is defined as behaviour that benefits another creature at the expense of the one carrying it out. In other words, altruism means doing something that costs time, effort, or resources, for the sake of someone else. This might mean providing food for another animal, giving a warning signal to protect others while putting yourself at risk, or fighting an enemy to save another animal from harm. Examples abound in nature, from the social insects whose lives revolve around the good of their community to rabbits that thump warnings of approaching footsteps, and vampire bats that share meals of blood. Humans are uniquely cooperative and spend a great deal of their time doing things that benefit others as well as themselves: what psychologists sometimes refer to as 'prosocial behaviour'. They have moral sensibilities and a strong sense of right and wrong. They are altruists.

Altruism is a problem for many social psychologists and economists who assume that humans rationally pursue their own interests. It is also a problem for Darwinism, although it was not always seen that way. The problem varies according to the level at which you think selection takes place – or, putting it another way – what you think evolution is for. If you believe, as many early Darwinians did, that evolution ultimately proceeds for the good of the individual, then why should any individual behave in such a way as to incur serious costs to itself while benefiting someone else? All individuals ought to be out for themselves alone, and nature ought truly to be 'red in tooth and claw'. Yet clearly it is not. Many animals live social and cooperative lives, parents lavish devotion on their offspring, and many mammals spend hours of every day grooming their friends and neighbours. Why do they do it?

An answer that does not work is what the British philosopher Helena

Cronin (1991) calls 'greater-goodism' – the view that evolution proceeds for the good of the group or the species. Greater-goodism permeated biological thinking in the early part of the twentieth century and is still a popular way of misunderstanding evolution. On this view selection works 'for the survival of the species' or 'for the good of humankind'. The reason this cannot work is simple. Think about the chance of infiltrators. Let us suppose there were a species of wild dog in which each dog gladly caught rabbits for every other dog, and the pack lived in amiable harmony. As long as this harmony prevailed all the dogs would benefit. But now imagine that a new dog appears that just eats all the meat he is given and never bothers to do any catching. He will, of course, get the best food, have more time to pursue the best bitches, and will generally live better. He will then, no doubt, pass on his selfishness genes to his many well-fed puppies. So much for the good of the pack – selfishness for the individual must pay.

The problems of thinking in terms of the good of the species were gradually recognised and since the early 1960s 'group selection' has been almost entirely abolished from neo-Darwinism (I shall consider some exceptions later). The answer that has so successfully transformed the problem of altruism is selfish gene theory. If you put the replicator at the heart of evolution and see selection as acting to the advantage of some genes rather than others, then many forms of altruism make perfect sense.

Take parental care, for example. Your own children inherit half of your genes. Your children are the only direct way your genes can be carried on into future generations and so parental care is obviously needed, but this same principle can be applied to many other kinds of altruism. Darwin hinted that 'selection may be applied to the family' (1859, p. 258) but did not pursue the idea any further. The British biologist J. B. S. Haldane first noted, in 1955, that a gene for selflessly jumping into a dangerous river to save a drowning child could flourish easily if that child were your own, and might still flourish, though less easily, if you saved your cousin, your niece or another more distant relative.

In 1963 a young PhD student in London, tackling on his own the unfashionable topic of altruism, had just had his first paper turned down. He became so lonely struggling with the unfamiliar mathematics involved that he sometimes used to work all evening in the main hall at Waterloo railway station just to have people around him (Hamilton 1996). But William Hamilton's next paper, 'The genetical evolution of social behaviour' (1964), became a classic. He put numbers to Haldane's suggestion and developed what has come to be known as the theory of kin selection. He imagined a gene G that tends to cause some kind of altruistic behaviour, and explained that 'Despite the principle of "survival of the fittest" the ultimate criterion which determines whether G will spread is not whether the behaviour is to the benefit of the behaver but whether it is to the benefit of the gene G.' (Hamilton 1963, p. 355). This means that altruistic behaviour can spread in a population if animals are altruistic towards their own kin. The nearness of the relationship determines just how much it is worth paying for the possibility of aiding the spread of the gene. Instead of basing everything on the idea of individual fitness, the important quantity becomes 'inclusive fitness', which takes into account all the indirect ways in which a gene can benefit (Hamilton 1964). The mathematics can get extremely complicated in real-life situations, but the principle is simple.

Genes are invisible. A monkey that is going to share some food cannot be sure whether the other monkey is really her sister or not, and certainly cannot look inside and find out just which genes the two of them have in common. However, this does not matter for the principle to work. Monkeys that, in general, share resources with kin more than with non-kin will get more of their genes into the next generation. How they achieve this may vary, and probably involves various simple heuristics such as 'share with another monkey you were brought up with' or 'share with other monkeys that look, smell, or feel like your mother' or 'share most with monkeys you spend most time with'. Depending on the lifestyle of the animals concerned, different heuristics will work better than others. They work not by making the monkeys calculate sums, but by giving them feelings that make them act appropriately. The same applies to us. In other words, people 'execute evolutionary logic not via conscious calculation, but by following their feelings, which were designed as logic executers' (Wright 1994, p. 190).

We humans love our children (most of the time) and however much we are annoyed by our brother or despise our aunt, we still find it natural and unsurprising that we give them birthday presents, send them cards, or care more about them than some person we met in the street. But the theory of kin selection explains far more of the detail of family dynamics than just that, including battles over weaning, siblings competing for their parents' resources, and other forms of family strife as well as love.

Another success for biology has been reciprocal altruism. Darwin (1871) speculated that if a man aided his fellow-men he might expect to get aid in return. A hundred years later Robert Trivers (1971) turned this speculation into the theory of reciprocal altruism, explaining how selection might favour animals who reciprocated friendship, for example, by sharing surplus resources in good times in the hope of help in bad times. Research has revealed that many animals do just this, but there is a catch. If you are going to pay back favours, and avoid being cheated, you must be able to recognise other individuals. Most animals cannot do this, but many primates can – as can elephants, dolphins, and even such unlikely species as vampire bats. Vampire bats have a special problem in that they are very small and can easily die if they go without a meal of blood for more than two nights in a row. Fortunately, blood meals can be much bigger than one bat really needs. So the answer is to share your blood – and keep a track on who owes what to whom.

Gratitude, friendship, sympathy, trust, indignation, and feelings of guilt and revenge have all been attributed to reciprocal altruism, as has moralistic aggression, or our tendency to get upset over unfairness. If we have evolved to share resources with other humans, but to make sure our genes benefit, then our feelings are the way evolution has equipped us to do it. On this theory not only moral sentiments, but ideas of justice and legal systems can be traced to the evolution of reciprocal altruism (Matt Ridley 1996; Wagstaff 1998; Wright 1994).

Game theory has made it possible to explore how and why various strategies might evolve. Trivers used a game called the Prisoner's Dilemma in which two people are kept apart and told they are accused of a crime with a penalty of, say, ten years in prison. If both stay silent they can be convicted only on a lesser charge and both get a shorter sentence, say three years, but if one gives evidence against the other the defector gets off free. What should they do? Obviously the best outcome all round is for both to stay silent – but there is a strong temptation to defect – and what if the other one is tempted? – you might as well be tempted too. There are many other versions using points, money, or other resources. The important point is that a perfectly rational and selfish person will always gain by defecting. So how does cooperative behaviour ever come about?

The answer is that in a one-off game it never should, but life is not a one-off game. We meet people again, and form judgements about their trustworthiness. The answer to the Prisoner's Dilemma lies in repetition. In iterated Prisoner's Dilemmas people assess the other's likely behaviour and then both can gain by cooperating. Players who have not met before often copy each other, cooperating with cooperators and not with defectors. Persistent defectors are shunned, and so lose their chance of exploiting others.

Games like this are also used by economists, mathematicians and computer modellers. In 1979, the American political scientist Robert

Axelrod set up a tournament and asked computer programmers to submit strategies for playing the game. The fourteen entries each played 200 times against all the others, themselves, and a random program. To many people's surprise, the winning program 'Tit-for-tat' was both simple and 'nice'. Tit-for-tat began by cooperating and then simply copied what the other player did. If the other player cooperated then both continued to cooperate and both did well; if the other player defected, Tit-for-tat retaliated and so did not lose out too badly against defectors. In a second tournament over sixty programs tried to beat Tit-for-tat but failed.

Subsequent research has used more complex situations, with many players, and has been used to simulate evolutionary processes. It turns out that unless Tit-for-tat begins against overwhelming numbers of defecting strategies, it will spread in a population and come to dominate it. It is what is known as an 'evolutionarily stable strategy'. However, the real world is more complex, and Tit-for-tat does not do so well when mistakes are made, or when there are more players and more uncertainty. Nevertheless, this approach shows how group advantage can emerge out of purely individual strategies without the need to appeal to evolution for the 'greater-good'.

Is this how cooperative behaviour actually evolved? If so it would need some kind of nice behaviour to get it started, and Trivers has suggested that kin selection might have provided the starting point. Animals already supplied with feelings of affection and caring towards kin could easily begin generalising and so give nice Tit-for-tat the start it needed.

Note that the Prisoner's Dilemma is a non-zero-sum game. In a 'zero-sum' game what I gain you lose, and vice versa. This is not so for many real-life situations. Half a blood meal means life or death to a hungry young vampire bat but no more than an easy way to buy future favours for a well-fed more experienced hunter. This exposes the rather unpleasant concept of bargain hunting – giving deliberately to others who are in great need because their debt to you will be all the greater. This approach has also been used to show how moralising might evolve, since it pays to punish defectors and even to punish people who fail to punish defectors. In this kind of game, trustworthiness becomes a valuable currency. It pays you to be seen to be cooperative because you may reap the reward at some later date.

I have given only a few examples of how sociobiology has dealt with the problem of altruism (more extended treatments can be found in Cronin 1991; Matt Ridley 1996; and Wright 1994) but I hope these are enough to see just how successful it has been. In a sense this approach takes the altruism out of altruism. Acts of kindness and cooperation can be explained because they ultimately help the survival of the selfish genes on which they depend. Is that the problem solved, then? Does all of human altruism ultimately come down to kin selection and reciprocal altruism?

The oddities of human altruism

In today's world we frequently deal with people who are unrelated to us and whom we know we will never meet again. This suggests that society ought to be becoming less kind and cooperative, but this does not seem to be happening. Psychologists have long studied helping and cooperative behaviour. Experiments in the 1970s concentrated on bystander apathy – the depressing finding that people often do nothing to help a person injured in the street. They found that helping is greatly increased if the bystander is the only one who can help, and is decreased if other people can be seen not helping – so this is another situation in which people imitate each other. More recent studies, however, show that people will offer help in a wide range of situations. Experiments teasing out the effects suggest that people help because they feel empathy for the sufferer, and not because they are related to them, nor because they can expect any reward for helping (Batson 1995).

Try to think of the most altruistic of human acts you can. Dawkins gave the example of giving blood. In Britain every healthy adult is encouraged (or at least invited) to give blood twice a year, and donors are not paid – you get a cup of tea and a biscuit, and a little badge after ten donations. He suggested this was a case of 'pure, disinterested altruism' (Dawkins 1976, p. 230). Others have suggested giving a large tip in a restaurant you will never visit again, or going to Ethiopia to help starving orphans. We might add picking up valuables found in the street and handing them in to the police, clearing away someone else's abandoned rubbish, recycling your waste, or setting up a standing order to a charity whose members you will never meet. Then there are dogs' and cats' homes, and many people who care for birds with broken wings or maltreated donkeys. All these may appear to be examples of 'true' altruism but, sociobiologists would argue, they are really the by-products of kin selection and reciprocal altruism. We are most generous to our relatives (or those we think might be relatives) and we are nice to others so as to build up a reputation for being good and trustworthy. Is this explanation adequate?

Let us take a few examples in more detail. Imagine an Australian who sends money to the starving in Africa, or an American who sends money to Bangladesh. Many people do this and some make no fuss about it. They send off a cheque and never even tell anyone they have done so. This cannot be kin selection because the final recipients are probably about as unrelated to the average donor as they could possibly be. You might even argue that on a planet with limited resources, this kind of generosity is strongly against the genetic interests of the donor – over and above the cost of the gift. So is it reciprocal altruism? Clearly not in any straightforward sense because the donor never expects to see the recipients or to be thanked by them in any way. However, evolutionary psychologists argue that such generosity is a way of building up the donor's reputation as a generous person (Matt Ridley 1996). In that case, though, we should expect people to brag about their donations, which often they do not. Even this can be explained as part of reciprocal altruism on the theory that the feeling of guilt is evolution's way of making sure the system works, and so these hidden acts of generosity are just mistakes – the price we pay for having our uniquely human emotions.

The examples I have given so far are mostly isolated acts of generosity, but altruism is much more deeply embedded in our lives than that. Vast numbers of people choose to do jobs that are badly paid, poorly rewarded, have very long hours, and are highly stressful, because they want to be of service. Such jobs include social work, psychotherapy, working in old people's homes, looking after delinquent children, and environmental protection. Why would anyone want to spend several years training to become a nurse and then spend their life working irregular hours, long shifts, dealing with difficult people, clearing up horrible messes, spending hours giving out pills and making beds in an environment of sickness and disease, all for an uncomfortably low salary? The answer cannot be for material gain or genetic advantage. Nurses may say it is because they want to help people, because it makes them feel fulfilled, because they believe that life is only worth living if you help others, because they are grateful to be healthy and want to help those who are not, because they recognise that money alone is not the way to happiness, and so on.

According to sociobiological theory these reasons must all be by-products of reciprocal altruism, but to me that stretches the theory to breaking point. The problem is that natural selection is ruthless and the cost of this kind of generosity could be very high indeed. People who managed to avoid paying it in the past would have been at an advantage and would have passed on their genes for avoiding it. Evolutionary psychologists might argue that our emotional system was designed for the hunter-gatherer lifestyle and must be expected to go wrong (and perhaps to produce excess generosity) in a rich technological world. Perhaps the knowledge that 'I will never ever see this person again' is no match for underlying emotions programmed by the genes in times past, but then we are back to explaining away our behaviour as just a mistake.

So is there an alternative?

Until now there have been only two major choices in accounting for altruism. The first is to say that all apparent altruism actually (even if remotely) comes back to advantage to the genes. On this view there is no 'true' altruism at all – or rather, what looks like true altruism is just the mistakes that natural selection has not managed to eradicate. That is the sociobiological explanation. The second has been to try to rescue 'true' altruism and propose some kind of extra something in human beings – a true morality, an independent moral conscience, a spiritual essence or a religious nature that somehow overcomes selfishness and the dictates of our genes; a view that finds little favour with most scientists who want to understand how human behaviour works without invoking magic. Neither choice appears satisfactory to me.

Memetics provides a third possibility. With a second replicator acting on human minds and brains the possibilities are expanded. We should expect to find behaviour that is in the interests of the memes, as well as behaviour serving the genes. Magic is no longer required to see why humans should differ from all other animals, nor why they should show far more cooperative and altruistic behaviour.

We can ask our meme-selection question again. *Imagine a world full of brains, and far more memes than can possibly find homes. Which memes are more likely to find a safe home and get passed on again?* I suggest that among the successful memes are altruistic, cooperative, and generous ways of behaving.

Altruism in the service of the memes

Imagine two people. Kevin is an altruist. He is kind, generous, and thoughtful. He gives good parties and buys people drinks in the bar. He often has friends round for meals and he sends out lots of birthday cards. If his friends are in need he takes the trouble to ring, to help them out, or to visit them in hospital. Gavin is mean and selfish. He resents buying other people drinks, and thinks birthday cards are a waste of money. He never invites people round for a meal, and if his (few) friends are in trouble he always has something more important to worry about. Now the question is – who will spread more memes?

Other things being equal, Kevin will. He has more friends and spends more time talking to them; they like him and they listen to him. The memes he spreads might include the stories he tells, the music he likes, the clothes he wears, and the fashions he follows. They might be the scientific ideas he likes to discuss, the economic theories he espouses, and his political views. Most important, they will also include all those memes that make him the way he is – memes for giving good parties, for sending out lots of cards, for helping people in need and for buying them drinks. Psychological experiments confirm that people are more likely to be influenced and persuaded by people they like (Cialdini 1994; Eagly and Chaiken 1984). So his friends will imitate his popular behaviour and thus his altruism will spread. And the more friends he has, the more people can potentially pick up his ways of making himself popular. We could call Kevin a meme-fountain (Dennett 1998).

Meanwhile, Gavin has few friends. He makes few opportunities for talking to the ones he does have, and he rarely finds himself chatting over a drink or passing the time of day with a neighbour. His memes have few chances to replicate because the few people who could potentially imitate him rarely do so. Whatever he thinks about the state of the nation or the best way of making apple pie, his ideas are unlikely to spread far because people do not listen to him, and if they do they do not adopt his ideas because they do not like him. We might call Gavin a meme-sink.

This difference forms the basis of a memetic theory of altruism. The essential memetic point is this – if people are altruistic they become popular, because they are popular they are copied, and because they are copied their memes spread more widely than the memes of not-so-altruistic people, *including the altruism memes themselves.* This provides a mechanism for spreading altruistic behaviour.

Note that I am not the first to treat altruistic acts as memes. As we shall see, Allison (1992) proposes quite a different mechanism, and Du Preez (1996) considers selfish and altruistic discourses to be evolving memes, though without explaining exactly why altruism should spread in spite of its cost. There are many ways of being altruistic and I have lumped them all together here, but they include generosity, kindness, caring behaviour, and so on – anything that makes it more likely that others will want to spend time with those people, and emulate them, and so will pick up their memes. Note that for this kind of memetic altruism to work two things must be true. First, that people are capable of imitation, and second, that they more often imitate altruists. If both these are true we should expect people just to find themselves being helpful and altruistic, without necessarily knowing why.

I am going to speculate about the origins of such behaviour in our evolutionary past. (In the next chapter I will deal with altruism in today's world where it is easier to test the consequences and to find out whether a memetic theory is really needed.) We begin with reciprocal altruism. People are nice to each other to get kindness in return, and their emotions are designed appropriately – that is, people want to be generous to those who might repay them, and they want to be liked. Now, add the capacity to imitate, and the strategy 'copy-the-altruist', and two consequences follow. First, kind and generous behaviours will spread by imitation. Second, behaviours that *look* like kind and generous ones, or are prevalent in kind and generous people, will also spread by imitation.

I speculated previously about how human imitation ever came about, and it is interesting to realise that Tit-for-tat entails a kind of imitation – it is essentially a strategy of 'imitate the other guy'. So perhaps the selection pressures favouring cooperative behaviour also played a part in the evolution of imitation itself. In any case, once imitation had arisen, people could begin copying each other, and ways of doing things could spread through whole populations. Among these ways of doing things would be acts of generosity, such as sharing food, giving presents, and looking after the sick – all of which could arise from sound genetic principles such as those we have already considered, kinship patterns, mating systems, and reciprocal altruism.

Once imitation has arisen, this process only works if people are more likely to copy altruists. This makes sense, because if you live in a community that uses reciprocal altruism you are likely to gain most by being with people who are known to be generous. So the generous people will have more contact with others and therefore more opportunities for spreading their memes. However, there is another reason why it might pay to copy altruists. A fundamental principle of reciprocal altruism is that people are most generous to people who are generous to them. But there is a way to cheat the system. If you want the rewards (other people's generosity) without paying the costs (actually being generous) you could try to *look* like a generous person. In other words, it would pay you to copy the people who really are generous. So the strategy 'copy-the-altruist' should spread. This strategy is, at first, of benefit to the genes but because it involves the second replicator the genes cannot keep it under control. 'Copy-the-altruist' starts as a strategy for biological gain, and ends up as a strategy for spreading memes – including (but not restricted to) memes for altruism itself. There will always be pressures *against* altruistic acts because of the costs involved, but once imitation is possible there is also memetic pressure *for* altruism.

Imagine two early hunters who go out with bows and arrows, leather quivers, and skin clothing, and both come back with meat. One, let us call him Kev, shares his meat widely with surrounding people. He does this because kin selection and reciprocal altruism have given him genes for at least some altruistic behaviour. Meanwhile Gav keeps his meat to himself and his own family, because his genes have made him somewhat less generous. Which behaviours are more likely to get copied? Kev's of course. He sees more people, these people like him, and they tend to copy him. So his style of quiver, his kind of clothing and his ways of behaving are more likely to be passed on than Gav's – including the altruistic behaviour itself. In this way Kev is the early equivalent of the meme-fountain, and he spreads memes because of his altruistic behaviour.

Note that there are two different things going on here. First, the altruistic behaviour serves to spread copies of itself. Second, it spreads copies of other memes from the altruistic person. This second possibility could produce odd results. As with biological evolution, accidents of history can have profound effects. So, if it just happened that in one particular group of our ancestors the generous people happened to have made specially natty blue-feathered arrows, then blue-feathered arrows would spread more widely than brown-feathered ones, and so on. Whatever the kind of memes we are talking about, they may be driven to increase by the altruism of their bearers.

There are also more complicated ways in which altruism could be spread memetically. The sociologist Paul Allison (1992) has suggested a number of 'beneficent rules' whose contents may ensure their own survival. They all take the general form of 'Be good to those who have a higher than average probability of being carriers of this norm'. This principle depends not on the strategy 'copy-the-altruist' but on 'copy-the-successful'. As Allison explains, suppose A follows one of these rules and helps B. B may now be more successful because of the help he has received. He is therefore more likely to be imitated and therefore to pass on the rule which made A help him in the first place. In this way the rule spreads itself.

This process only works if B actually takes on the beneficent rule and does not just take the kindness and run. That is why the general rule is to be good to others who are likely to carry the rule. So who are they? Among many versions of the rules are 'Be good to those who imitate you', 'Be good to children', 'Be good to your cultural ancestors' or, more generally, 'Be good to your close cultural relatives'. For example, you might follow the rule 'Be good to your cultural descendants'. If people have already taken on other memes of yours and in general are known to copy you, then they are more likely to take on your beneficent rules as well. Since kindness to them is likely to increase their cultural fitness they may also pass it on to someone else and so the rule will thrive. This process would apply to biological parents and their children, in which case it would be hard to distinguish from kin selection. It becomes more interesting when applied to non-kin, and Allison considers the example of professors and their graduate students. Professors who are generous to their students (in time, effort, and so on) increase the cultural fitness of their students and hence the chances that all their memes, including the beneficent rule itself, will be passed on to yet more students. This makes sense, because a caring professor who works hard for her students' welfare will certainly attract more students – and better students – who in turn are likely to do the same.

Note that it is the rule that benefits here, not the professor. Perhaps rationally the professor should not be so generous, but because these norms thrive and she has picked them up, she will be. Allison does not use the term 'meme' but his beneficent norms clearly are memes, for he specifies that they are passed on by imitation and teaching. His analysis shows how taking the meme's eye view (or 'norm's eye view') can explain behaviours that cannot easily be explained in terms of rational choice theory or genetic advantage.

Note that Allison's scheme best accounts for acts of altruism directed towards cultural relatives and, as he points out, it cannot account for altruism directed at large groups of people, or at people in general. In contrast, memetic altruism based on 'copy-the-altruist' can explain just this kind of generalised altruism.

Memes versus genes

Any act of meme-driven altruism potentially lowers the actor's genetic fitness. In other words, the arena of human altruism can be seen as a competition between memes and genes. Kev's behaviour will make him friends but it may reduce his chances of survival, or the chance of his children's survival, by reducing their share of the meat. His genes only 'care' about his generosity if it serves in the long run to pass them on, and they have equipped him with feelings and behaviours that generally serve their interests. But his memes do not 'care' about his genes at all. If they can get copied they will. And they will, because people copy people they like. Thus we can imagine a human society in which meme-driven altruistic behaviour could spread – even if it put a heavy burden on individuals. In other words, once people start to copy the altruists, the genes will not necessarily be able to stop them.

Could memetic altruism get completely out of hand – and stretch the leash to breaking point? Sometimes people do give more than they can really afford. They vie with each other to be the most generous, or give the most ostentatious of gifts. As Matt Ridley (1996) points out, gifts can become bargains, bribes, and weapons. Most extraordinary is the practice of 'potlatch'. The term comes from the Chinook language and potlatch is best known from American Indian groups, but it also occurs in New Guinea and other places. A potlatch is a special event in which opposing groups try to impress their rivals by giving away, or destroying, extravagant gifts. They may give each other canoes and animal skins, beads and copper plates, blankets and food. They may even burn their most valuable possessions, kill their slaves, and pour precious oil onto a huge fire.

Note that this wasteful tradition is not like ordinary reciprocal altruism. In most forms of reciprocal altruism, both parties benefit from cooperating, but in a potlatch everyone loses (at least in purely material terms). Note also that potlatch depends upon imitation. Such a tradition could only spread by one person copying it from another until it becomes the norm for a whole society. It is imitation that makes such peculiar behaviour possible, and once the genes have given us imitation they cannot take it back. We could see the potlatch behaviour as like a parasite that may, or may not, kill its host, while most of our altruistic behaviour is symbiotic or even beneficial.

Once again we can see that it is our capacity to imitate that makes humans so different from other species. In other species gifts are confined to sharing with kin, to precise reciprocal deals, or to special situations such as the male spider who gives his mate a well wrapped fly to keep her busy while he copulates. Among human cultures, giving gifts is common; visitors bring gifts, special occasions are celebrated with gifts, marriages and birthdays are marked with gifts. In Britain, about seven to eight per cent of the economy is devoted to producing articles that will be given away as gifts, and in Japan the figure may be higher still. Fortunately, potlatches are rare, and for most of us the giving and receiving of gifts is an enjoyable part of being human.

One last step gives us meme–gene coevolution again. I have already argued that the best imitators, or the possessors of the best memes, will have a survival advantage, as will the people who mate with them. So the strategy 'mate-with-the-best-imitator' spreads. In practice, this means mating with those people who have the most fashionable (and not just the most useful) memes, and we can now see that altruism is one of the factors that determines which memes come to be fashionable.

So Kev, the meme-fountain, will not only make more friends and spread more memes, but since these memes are fashionable he will also attract a better mate and pass on the genes that made him altruistic in the first place. This means that insofar as the originally altruistic behaviour depended on genetic differences, these will be passed on to more offspring, and so altruistic behaviour will spread genetically, as well as memetically. Note that this process entails the memetic-driving of *genes* for altruism rather than just memes driving *memes* for altruism as described above. By this process genes for human altruism could have been meme-driven – making us genetically more altruistic than we would otherwise be.

Note also that this possibility arises because two strategies coincide – 'imitate the altruist' and (because altruism memes are imitated and become fashionable) 'mate-with-the-altruist'. The same does not apply for Allison's beneficent norms because they depend on the strategy 'mate-with-the-successful', which is directly in the interests of the genes, and is in any case widespread. In other words, for Allison's rules the outcome will be similar whether just genes were involved, or memes as well.

I suggested that 'imitate-the-altruist' had two consequences: spreading altruism memes and spreading other memes associated with altruists. The same applies to the memetic driving of genes. So, not only might genes for altruism be favoured but, by the quirks of history, other genes might be affected. For example, let us suppose that there was some genetic components to Kev's choice of blue feathers (differences in colour vision, for example). Blue-feathered arrows became popular because they first appeared on Kev, and Kev was a generous person. Now people not only copy the feathers, but they preferentially mate with people who have the fashionable blue-feathered arrows. Thus, the genes for preferring blue feathers may now have an advantage, and, if the fashion were maintained for enough generations, gene frequencies might start to change. Note that there need be nothing intrinsically better about having blue-feathered arrows. The whole process began only because it was an altruistic person who started the fashion.

I have no idea whether memetic-driving of this kind has ever taken place or not. There is some observational evidence that human infants show a tendency to share (as well as to be selfish of course) at a young age, while infants of other primate species do not, suggesting an innate basis. Certainly, humans have a far more cooperative society than any other species, apart from the social insects such as ants and bees that operate by kin selection. This theory of memetic altruism could provide the explanation. It might also help explain why the relationship between memes and genes is apparently so successful, even though the two replicators are so often at odds. Perhaps memes are more like a symbiont and less like a parasite precisely because they encourage people to cooperate with each other.

If there were many other species with memes, comparisons would be easy; but there are not. Many birds imitate each others' songs and so perhaps we should expect these birds to show more altruism to each other than closely related non-imitators. Dolphins are among the very few other species capable of imitation, and they are renowned for stories of heroic rescues. Dolphins have been reported to push a drowning human up to the surface of the sea, and even to push someone onto land – a very strange thing for another species to do. But this is only anecdote; much research would be needed to find out whether the idea is valid or not. Other research to find out whether memetic driving of altruism has ever occurred would be difficult, as is all research on behaviour in our distant past.

The prospects for research are much brighter when it comes to modern humans and their behaviour, and I want therefore to leave speculation about Kev and Gav and return to their modern counterparts. We shall see that being kind, generous, and friendly plays an important role in spreading memes in today's complicated society.

CHAPTER 13

The altruism trick

In today's world I am going to assume that we can ignore meme–gene coevolution. This must be an oversimplification, because as long as there are two replicators they will interact with each other. However, the pace of memetic evolution is now so fast, relative to that of human genetic evolution, that we can safely ignore the latter for most purposes. The genes cannot keep up. What we cannot ignore is the legacy left by the long process of coevolution. The brains we have are the big and clever brains created by meme–gene coevolution. The way we think and feel is a product of that evolutionary process, and now determines which memes do well and which do not. We like sex, so sex memes get a head start: different ones for men and for women. We like food and we like power and excitement. We find maths hard, and so mathematical memes need a lot of encouragement. The structure of our language affects which memes are more easily passed on. The theories and myths we have created affect the way we deal with new memes. And so on.

Note that sociobiology has made a different simplifying assumption and has ignored the role of the memes. For many purposes this has been an adequate approach, and we can use many of the findings of sociobiology to provide insight into the brains we have and the ideas and behaviours that come easily, but it cannot provide the whole picture. Our concern now will be what happens when vast numbers of memes compete to get into, and stay in, limited numbers of increasingly educated and overworked brains.

We must resume the meme's eye view; remembering that all that counts in the life of a meme is whether or not it survives and replicates. I shall find myself saying that memes 'want', 'need', or 'try to do' something. But we must remember that this is only shorthand for saying that the 'something' will improve the chances of the meme's being copied. Memes do not have conscious intentions; nor do they actually strive to do anything at all. They are simply (by definition) capable of being copied, and all their apparent striving and intentionality comes from this. When anything can be copied it can end up having few or many copies made. Memes may be successfully copied because they are good, true, useful or beautiful – but they may be successful for other reasons too. It is those reasons I now wish to explore.

A meme that gets into a meme-fountain will do better than one that only gets into meme-sinks. We can guess who the meme-fountains are. Indeed, many experiments in social psychology show who is most often emulated. Powerful people (and people who dress in the trappings of power), people perceived as experts, and people in authority are all examples of 'imitate-the-successful'. All these people are more likely to get others to do what they say or to accept their ideas; as salesmen, advertisers and politicians have long known. In discussing the 'power button' Brodie (1996) suggests that TV shows use large cars, guns, and flashy clothes to gain more air time and so promote their kind of memes. Fame spreads memes, as when television and film stars are watched by millions of viewers, so changing the fashions in clothes, speech, smoking or drinking, cars, food and lifestyle. But not everyone is powerful, and there are other kinds of meme-fountain. For example, we are more likely to be persuaded by someone we perceive as similar to ourselves, and a clever sales trick is to mirror the actions of the potential buyer or to pretend to having similar beliefs or hobbies (Cialdini 1994).

I have already suggested that one way to spread memes is to behave altruistically, and I now want to consider some of the consequences of this less obvious way of becoming a meme-fountain. First, altruistic behaviour spreads copies of itself – so making us more altruistic. Second, altruism helps to spread other memes – so providing a trick that memes can use to get themselves copied.

Altruism spreads altruism

Let us consider first the copying of altruistic behaviour itself. Imagine two different memes (or sets of memes). One is a set of memes for helping your friend when she is in trouble – whether it is giving her a lift when her car breaks down or listening to her troubles when her boyfriend leaves her. The other is a set of memes for ignoring what your friend needs. These are behaviours that can be copied from one person to another and so they must be memes. Note that I use the phrase 'a meme *for* something'. This is potentially dangerous because it might be taken to imply that there is a particular instruction explicitly stored somewhere in a brain which tells the person to help their friend – and this can easily be made to look ridiculous. This interpretation is not necessary, however. All that is necessary is to assume that people imitate aspects of each other's behaviour and that when they do so something is passed on from one to the other. We do not need to agonise about what that something is. The simple fact is that if imitation happens (as it surely does) then something has been passed on and that something is what we call the meme. So when I say a 'meme *for* helping your friend' I only mean that some aspect of helping behaviour has been passed on by one person copying another.

Now we can ask the important question: which of these two memes will do better? The first meme will – it will make your friend like you more and want to spend more time with you. She will then tend to imitate you more than her other, less helpful, friend and so your helpfulness memes will spread to her. She will therefore become more helpful to her other friends, and so the meme will gradually spread. The same simple logic applies to any meme which helps its carrier to become more popular. The people who pick up these memes are not aware of what they are doing, they just find themselves wanting to be more like the nice people, not the nasty ones. They find they want to help and be kind and feel bad if they do not. Just as many of our human emotions serve the genes, so these ones serve the memes – and they are no less noble for that.

Does this mean that everyone will become nicer and nicer and nicer without limit? Of course not. The main reason why not is that being kind and generous and altruistic is expensive in terms of time and money. There are always pressures acting against altruism, and there are always other strategies for memes to use. However, in general it means that people will be more altruistic than they would be if they were incapable of imitation.

This is an example of meme-driven altruism in a modern context (and note that this is different from the memetic driving of *genes* for altruism which I considered at the end of the previous chapter). In this kind of meme-driven altruism, actions that are costly and done for someone else come about through memetic competition. Because these actions are driven by memes and not genes they need not necessarily be in the person's genetic interest. These cases, in which the genes do not benefit and the memes do, provide test cases for a memetic explanation. People who devote their entire lives to charitable work or to the caring professions while having no children of their own are examples. Their sacrifice cannot easily be explained in terms of genetic advantage, but can be simply explained in terms of memetics.

In principle, meme-driven altruism ought to be able to produce the most pure and selfless generosity. Indeed, it may occasionally do so. However, altruism not only works to spread itself but also acts to spread other memes as well. This provides a mechanism open to exploitation by other memes. This, I suggest, is exactly what happens. I shall describe several ways in which memes can exploit the process of meme-driven altruism. These are all versions of what I shall call the 'altruism trick'.

The altruism trick depends on the simple idea that a meme that gets into an altruistic or likeable person (like Kevin) is more likely to be copied than one that gets into a meany (like Gavin). So what kinds of meme (other than memes for altruism) can get into the altruist?

First, some memes *look* like altruism even if they are not, and so they can fit easily in an already altruistic person, and second, memes can group together into memeplexes that use various tricks to get into altruists.

Looking like altruism

The first is an obvious trick, to *look* like altruism. A meme that makes a person appear to be kinder and more generous will increase the chances of that person being imitated and so of that meme being spread, without incurring great costs. There are many examples of this kind of behaviour. We smile at people a lot, and we smile back at people who smile at us first. We say kind and polite things to them – 'How are you?' 'I do hope your parents are well' 'Have a nice time at the party' 'How may I help you?' 'Have a good day' 'Happy New Year'. With all these common memes we give the impression of caring about the other person, even if we do not. That is why they are successful memes. Our ordinary everyday conversation is full of such memes.

Closely related to this is the sort of meme that sneaks easily into an altruist. Memes do not exist in isolation. All memes, at least at some phases of their lives, are stored in human brains, and humans are complicated creatures who strive to maintain some kind of consistency to their ideas. This 'consistency principle' is crucial in understanding a lot of human thought and action. If a given person tends to be altruistic, whether because of a genetic tendency to act that way, or because he has picked up lots of altruistic memes during his lifetime (or most likely because of both), then other altruism memes are more likely to gain a foothold there.

Let us suppose a new meme comes along in the lives of Kevin and Gavin; suppose they both hear a plea to save their used stamps and send them to some charity. This new meme is far more likely to be accepted and acted on by Kevin than Gavin. It fits well with his other behaviour. He thinks of himself as a caring person and so on. If he refused to take part he would suffer 'cognitive dissonance', the unpleasant consequence of holding two incompatible views – in this case, his idea of himself as a caring person and his refusal to help with the stamps. Many psychological studies have shown that people will work to reduce the dissonance between incompatible ideas, and also that consistency itself is generally admired and emulated (Cialdini 1994; Festinger 1957). The idea is less likely to take hold of Gavin. He would suffer no cognitive dissonance by refusing to help in this or any other way.

The need for consistency and the avoidance of dissonance provide the context in which memes club together in different people. Once someone is committed to a particular set of memes, other memes are more or less likely to find a safe home in that person's repertoire of arguments, beliefs, and behaviours. We find this kind of generalisation of memes in all sorts of contexts. You might think it is just common sense that nice people do nice things and nasty people do nasty things but memetics puts this common-sense fact in a slightly different light. Memes can succeed or fail because of the genetic propensities of the people they come across, but also because of the memes that are already present in those people.

The situation is all the more complex because of changing fashions. The memes which are acceptable will shift as the whole meme pool changes. At one time, certain types of charitable giving will seem appropriate, but a few years later, completely different kinds will take over. But this complexity should not cloud the basic principle. Once meme-driven altruism has got going it will generalise. Memes for all sorts of kind and generous acts can take hold more easily in people who are already infected with altruistic memes and who have invested in a particular view of themselves. These people are copied more than other people and so these memes spread more widely.

This process can be used to understand all sorts of otherwise rather baffling actions. Let us take kindness to animals. Many people go out of their way to help animals in distress. There are homes for dogs and cats, and refuges for sick donkeys and injured wildlife. There are game parks and great international attempts to save species from extinction. There are 'Save the Animals' charity shops, and greetings cards that support wildlife organisations.

I say this is baffling because there is no easy explanation of all this inter-species kindness in terms of rational self-interest, genetic advantage, or evolutionary psychology. Rescuing an injured tiger would not benefit a hunter-gatherer. Animals were not domesticated until about ten thousand years ago in the 'Fertile Crescent' to the east of the Mediterranean, as recently as one thousand years ago in America, and not at all in some parts of the world (Diamond 1997). Therefore during most of our evolutionary past, he animals around us have mostly been either potential prey for eating or predators trying to eat us. Saving them from death makes no genetic sense; nor does working to relieve their suffering. I have never come across a sociobiological explanation of kindness to animals, although I can think of several possibilities. Animals cannot, on the whole, pay back the favours; so direct reciprocal altruism is no explanation. However, a possible argument is that reciprocal altruism has given us the emotions that drive this behaviour. We feel empathy with suffering animals and want to relieve it; we feel guilt if we do not, and so on. Another possibility is that we raise our status in the reciprocal altruism stakes by appearing so kind. I am not convinced that this makes sense, because of the high potential costs of such behaviour. Surely, natural selection would have weeded out any tendencies to be too kind to animals, especially wild and dangerous ones. These theories are also hard to test.

Why do we do it then? I suggest that kindness to animals can easily take hold because it fits well in people who are already infected with altruism memes. They see themselves as kind people and have an investment in continuing to be so. The way they behave makes them more likely to be imitated, and so kindness to animals spreads.

Exactly the same argument applies to the increasingly widespread practice of refusing to eat meat. Humans were clearly designed to eat a certain amount of meat. Meat is high in protein and fat, and was probably necessary to feed the increasingly large brain of our far ancestors. Yet now many people, myself included, do not eat meat. Some argue that they feel better on a vegetarian diet and a few do not like meat, but most say they are affected by the suffering of the animals bred and killed for food. I suggest that vegetarianism succeeds as a meme because we all want to be like the *nice* people who care about animals, and we copy them. Not everyone will get infected by this meme; some like meat too much and others have sets of memes that are not very compatible with this one. Nevertheless, it does quite well. Vegetarianism is a memetically spread altruistic fashion.

If this is right we should expect to be able to trace the historical origins of such memes as they gradually appear and take hold of whole populations. We would not expect to find such actions in societies with little communication and few ways for memes to spread. We would expect them to be most common in societies in which people have plenty of resources to spare and plenty of opportunities for picking up new memes. We should not necessarily expect people to brag about being kind to animals, but simply to find themselves wanting to be so.

Note that it is not necessary that the superficially kind actions should actually help the animals in question. An injured animal that is rescued is helped in the short term, and a potential battery hen that is never hatched is almost certainly better off for never having existed. But the long-term prospects are dubious, especially when it comes to schemes for saving whole habitats or species. The memetic approach makes it easy to understand why particular behaviours spread even when they do not achieve what they are supposed to achieve. It is not just that people make mistakes in their reasoning, which we know all too well, but that they are especially likely to make certain sorts of mistakes – in this case copying behaviours that *look* altruistic.

A final example of this kind is recycling waste. Recycling is certainly a meme – that is, a behaviour that people pick up by copying other people, whether they read about it, see it on television or discover that all their neighbours are doing it. Many people put a great deal of effort into separating different kinds of waste, storing them in their house or garage, taking them to recycling points, and buying recyclable goods. The recycling meme has been an enormously successful one, spreading far and wide in the developed world and driving a massive amount of human activity. Some experts argue that the energy thus used is far more than would be needed if the materials were simply dumped and new ones made. I have no idea whether this is true, but from the memetic point of view it does not matter. We would expect these kinds of behaviour to spread because they are easily picked up by people who already do all kinds of generous, caring and 'green' activities, who are therefore seen as altruistic and are therefore copied. The whole 'green movement', and the effort put into it, is just what you would expect of meme-driven altruism in action.

Memeplexes and the altruism trick

Memes which have nothing to do with altruism can benefit from 'copy-the-altruist' by just tagging along for free. Like Kev the caveman's flashy blue-feathered arrows, some memes may just by luck happen to be carried by more altruistic people, but this luck is not a memetic process that can be relied on. Instead, we can expect memes to have devised strategies for getting into altruistic people without actually being altruism memes themselves (or more accurately – memes that happened to have such strategies should have survived better than those without, and we should be able to observe them around us). Are there such examples?

Yes. They range from little groups of co-memes to very complicated memeplexes. Remember that the essence of any memeplex is that the memes inside it can replicate better as part of the group than they can on their own. Some simple ones will show the principle. For the first type we need to assume that people want to be liked. This is part of the principle I have been following that people imitate people they like more than people they do not. Imitating people you like should be a good way to become liked yourself and being liked should ensure that people are nicer to you.

Now, let us take some actions a parent might try to persuade a child to do, such as keep clean, say please and thank you to Auntie Dawn, or stay a virgin until after marriage. Why should children obey the instructions? They might obey out of fear or coercion, but a common trick is to turn the instruction into 'Good children keep their clothes clean', 'Nice people say please and thank you', or 'Good girls don't have sex before marriage'. These simple memeplexes consist of just two parts; the instruction and the idea of being good. 'People won't like you if you do that' is another, as are hints that nice people vote conservative, people like us eat dinner at eight, or kind people go to church.

More complicated memeplexes can build up around the kinds of altruism I considered before, such as kindness to animals or recycling, and lots of other memes can jump on board. The recycling symbol is a little scrap of information that has been very successfully copied around the world. The names and logos of all the charities are other examples, as are the collecting boxes that are rattled in the street, the practices of having charity shops, of distributing special bags to collect goods in, and many other activities that thrive in the world of charitable giving. As memeplexes evolve and become more complicated, new niches are created in which new kinds of meme can thrive. In the examples I have given here, the spread of charitable giving opens up niches for all sorts of other memes to thrive.

You can even sell music and fashions using altruism. Bob Geldof really did give money to the starving in Africa but he sold millions of records at the same time. Princess Diana's memorial fund really is funding her charities but it is spreading millions of Diana memes in the process – pictures, stories, personal reminiscences, speculations and scandals, videos of her life and times, not to mention the words and tune of Candle in the Wind.

These are simple examples, but they are sufficient to show that meme-driven altruism is an obvious meme-trick ready for exploitation. It should not, therefore, surprise us to find that many of the most powerful and widespread memeplexes use it in various forms. Pre-eminent are the religions. One of the mechanisms is simple, once you think about it memetically. A religion which persuades its followers to be more altruistic will spread because of the altruism trick.

I was once cycling in a park in Bristol when my bicycle chain fell off. Before I could jump off to put it back two young men raced up to me, politely offered help, expertly put the chain back on, and stood smiling kindly at me. 'Thank you very much', I said, feeling a little bewildered. For I had never seen them before and I was not a ravishing sight in my Felix-the-cat bike helmet. God was soon on their lips, quickly followed by Joseph Smith and Salt Lake City. The Mormon faith is ably and deliberately spread by the altruism trick. It doesn't work on everyone, but it works well enough to keep the memes alive.

The altruism trick works like this. Take a political party, a religious sect, a cult, a local benevolent society, or any complex belief system. Incorporate within it the idea that its followers should do good works. These good works will then make the followers more likeable and so people will copy them – copying in the process all the other memes in the belief system. Of course, this mechanism does involve actual 'good works', as did Geldof and Diana. Others only give the appearance of doing good, or just persuade their followers to *think* they are doing good. Others exploit the sense of obligation induced by giving gifts – the proselyte does you a good turn, you now feel obligated to him, and the obvious way to repay this obligation is to do what he wants, that is, to take on his memes (or at least give the appearance of doing so). There are many variations on this basic 'altruism trick'. I will consider how some of them work, as well as further implications of Allison's (1992) beneficent norms, when dealing in more detail with religions.

Note that this trick effectively makes people work for the memes they carry. People who join the cults or adopt the ideologies give away their possessions, do good works, or help others, because this helps copy the memes that have infected them. Other people then copy them and they also begin to work for the memes. This is one reason why memeplexes that use this trick have survived in the past and why there are so many of them around now. This is the second time we have met the idea of people working for their memes (the first was in relation to sex and spreading memes rather than genes) and we will meet it again. In this sense we can say that the memes are driving human behaviour.

If this seems frightening then we need to ask ourselves why. What does drive human behaviour? Much of the antagonism towards Darwinism, sociobiology, and indeed any science of human behaviour, stems from an apparent desire to see ourselves as magical autonomous agents in charge of our own destinies. I shall tackle the basis of this view later, but for now just say that yes, memetics does undermine this view. We can describe any behaviour in numerous different ways for different purposes, but underneath them all lies the competition between the replicators. Memes provide the driving force behind what we do, and the tools with which we do it. Just as the design of our bodies can be understood only in terms of natural selection, so the design of our minds can be understood only in terms of memetic selection.

Debts, obligations and bartering

Can the theory of memetic altruism be tested? One approach would be to test the basic assumptions on which it rests. The main assumption is that people preferentially copy the people they like. I have assumed this because there are substantial hints in the literature that this is so. In his widely cited book on the psychology of persuasion, the American psychologist, Robert Cialdini (1994) reviews the evidence that people are more easily influenced by, and more likely to agree to a request or buy a product from people they like. Tupperware parties work because the hostess invites friends who like her and are therefore more likely to buy products they do not want. Successful car dealers charm their intended purchasers by complimenting them, appearing to be similar to them, giving away small concessions or appearing to take their part against the boss, all of which increases the clients' liking for the dealers and hence the ease with which the victims can be separated from their money. The major factors that increase liking include physical attractiveness, similarity, cooperativeness, and the belief that the other person likes you. One record-breaking salesman even used to send out thirteen thousand cards a month to his clients saying 'I like you' – and presumably he was not wasting his money.

What is not so clear is whether liking leads directly to imitation. This has not been much studied by social psychologists, perhaps because the importance of imitation *per se* has not been emphasised. If it does, the other consequences should follow; that people buy products from, are persuaded to change their minds by, and more often agree with people they like. In other words, the social psychological findings described above may be a consequence of a deeper underlying tendency to want to copy people we like. The experiments that need to be done, therefore, should look more closely at the imitation of actions carried out by likeable and unlikeable people. For example, we might ask people to watch liked and disliked models carrying out a task in different ways, and then do the task themselves. Experiments could then go on to find out just how best to manipulate liking so as to produce the most effective imitation. If the same manipulations affect simple imitation of actions as well as persuasion and agreement with beliefs, this would be suggestive that a similar process is going on in both. I have also assumed that altruistic behaviour makes people more likeable. This may seem too obvious to need testing, but we could use similar experiments to test the main consequence of this – that is, that acting altruistically will induce people to imitate you. If these predictions were not born out the entire basis of this kind of meme-drive altruism would be undermined.

The outcome of such experiments might be complicated by the effects of the 'reciprocation rule'. It is well known in social psychology that people feel obliged to repay any kindness shown to them, and feel obligated if they do not (Cialdini 1995). This tendency is culturally widespread and probably related to the fact that aid from rich to poor countries is not always well received (Moghaddam *et al.* 1993). Presumably, reciprocity stems from our evolved use of reciprocal altruism. Now, if an observer in one of our experiments has a kindness done to them they may feel obligated to the model – an unpleasant feeling which might disincline them to like the model and so complicate the issue. The most interesting outcome from the memetic point of view would be if imitating the altruist (i.e. taking on their memes) acted as a kind of reciprocation. By this I mean that one person could 'pay back' a kindness by taking on the other person's ideas.

This effect can be seen to follow from a combination of the 'reciprocation rule' which derives from reciprocal altruism, and Allison's beneficent norm 'Be good to those who imitate you'. According to this rule, if A imitates B, B should now feel obliged to A. So, for example, not only does the professor want to be nice to her students but all of us should be kinder to people who agree with us, or take on our ideas, or imitate us in other ways. If the process works both ways then if C gives D a gift, D will feel obligated to C and may pay back the obligation by agreeing with C (or taking on her memes in some other way). In ordinary life we may be seeing this in the tendency of guests to agree with their host's ideas, or of people in subordinate positions to agree with those who have power over them, or in the tricks used by religions that I discussed above. Finally, this could lead to people trading off their obligations by bartering goods against imitation in all possible combinations. So, for example, the guest who brings a fine present should feel under less obligation to agree with the host than one who does not.

If the idea of exchanging goods for taking on memes seems unfamiliar, we might think of the bartering of memes that goes on all around us. We are used to the idea of paying for the information we want, by buying books or newspapers, paying our TV licence, or buying tickets to the cinema, but if people want to impose their ideas on us then they have to pay to get our attention, like advertisers and politicians do. I shall return to this in considering the way that information is put onto the Internet at the cost of the provider, not the user.

All these exchanges could be investigated. Imagine an experiment in which James expresses some unpopular idea, or solicits people to join his organisation, or whatever. Among a group of people present, Greg gets up and publicly agrees with James. Now James should feel obliged to Greg and so be more likely to act generously towards him than to the others. Such experiments could find out whether exchanging memes could become a kind of currency like exchanging goods.

Other experiments might bring together people of opposed viewpoints, or people who disagree about the right way to do something, and find out what methods they actually use to change one another's minds. Studies of attitude change have often been done where material gain is at stake, such as in advertising and political persuasion, but this theory predicts that people will, if given the chance, be more generous to people they are seeking to convince – even when there is no material gain involved. In addition, there is no point being generous to people who already agree with you, nor to those whom you judge as being beyond conversion. The greatest altruism should be shown to those who are capable of being convinced (Rose 1997).

The effects of reciprocation are a little more complex, however. Imagine the following experiment. Just two people are involved (though in practice we would need to repeat it with many pairs). Janet is asked to express her opinion on some controversial topic while Meg listens in silence. Janet now acts generously in some way towards Meg (perhaps by buying her a coffee or offering to help with something). Meg is then asked to say how much she likes Janet. We should obviously expect that Meg will express greater liking for Janet when she has been generous towards her than when she has not. Now we give Meg the chance to say what *she* thinks about the controversial topic and again measure her liking for Janet. The theory makes two predictions. The first is more obvious, that Meg is more likely to express agreement when Janet has given her something. The second is less so, that expressing agreement acts as a kind of repayment of the kindness. So we should predict that if Meg now publicly agrees with Janet (whether that is really her opinion or not) she will now like Janet *more* than if she does not. In other words, Meg likes Janet not only because she was kind to Meg, but because Meg has paid off her debt by agreeing and so need no longer feel under any obligation to Janet.

This is an extremely artificial situation but I have tried to keep it simple. More realistic ways of taking on someone's memes might be to copy their actions in some more concrete way, to agree to pass on information to someone else, to write down what they say, to join a group they belong to, and so on, but I hope the principle is clear – that liking for a generous model would be *increased* if the subject were given the chance to imitate her, because the sense of obligation was reduced. This is, I suggest, a counter-intuitive outcome that could not readily be predicted or explained on any other theory.

If these predictions are correct they suggest that memes and resources can be bartered against each other in all sorts of ways. We should be able to pay people to accept our ideas, agree with people to pay off debts, and force people into agreement by what appear to be generous actions. There are interesting implications here for the power of money to coerce people into agreement. Some of the predictions are fundamental to the processes underlying meme-driven altruism and therefore, if they do not work out, my theory is wrong.

CHAPTER 14

Memes of the New Age

One day in 1997, a young student came to interview me for his media project. After several predictable questions he said 'Dr Blackmore, you are well known for your theory that alien abductions are really a form of sleep paralysis. Well I have experienced sleep paralysis *and* I have been abducted by aliens, and I can tell you they are not the same thing at all.'

It was my turn to ask the questions. Over many hours he told me of multiple abductions, starting when he was five years old and continuing into his adulthood. He told of aliens landing in the fields outside his house, of their visits to his bedroom and of operations they performed on him inside their spaceship. To top it all he showed me a tiny metallic object that the alien creatures had implanted in the roof of his mouth and which he had removed after two weeks of discomfort. Would I, with my 'closed mind' on UFOs, be prepared to analyse it scientifically?

Naturally I said yes. My own sceptical view of abductions was open to test by just this kind of object. There have been thousands of claims that people have been abducted, and several well-known academics prepared to support them (Jacobs 1993; Mack 1994). The stories are fairly consistent, and the people who tell them are known to be of at least average intelligence and education, and generally psychologically healthy (Spanos *et al.* 1993). But absolutely no convincing physical evidence has ever been provided, unless you count some stained clothing and a few previous 'implants'. But you never know, this one might be it – every scientist's dream – an object of unimaginable consequence – a piece of technology from an alien civilisation. Of course I wanted to analyse it.

The analysis was simple and so was the answer. The mysterious object, though it looked very much like other 'implants' under the electron microscope, turned out to be made of dental amalgam. The young man was partly disappointed and partly relieved, but as far as I know he is still convinced that he has been abducted, even though he is not now so frightened of the creatures implanting more objects in his body.

So what is going on? Accusing these people of either making up their incredible stories, or suffering delusions, is unfair when many of them (and I have met many) appear perfectly ordinary, sane people. They have clearly been frightened by something that has happened to them, and are convinced that aliens are to blame.

Alien abduction as a memeplex

I suggest that the aliens are a memeplex; a memeplex consisting of the idea of four-foot high skinny, large-headed creatures with big black eyes, an image of the ships they come in and the operations they perform, their intentions in visiting our planet, and all the other details we are fed through the media. As Elaine Showalter argues in her book *Hystories* (1997), such epidemics are spread by stories (though I would not describe them all as *hysterical*). Interestingly, the alien intentions vary with the group you favour. Followers of John Mack are inclined to the ecologically friendly type of alien who is coming to warn us of impending catastrophe, while followers of the Jacobs' school are abducted as part of an alien breeding programme making half-human-half-alien babies to invade our planet.

The memetic approach to alien abductions is to ask why these ideas should propagate so successfully, when they are not true. There is no mystery about why true and useful ideas should propagate successfully – they do so because people want and can use them. So memetics does not provide much advantage over other ways of looking at the world when it comes to understanding the success of good scientific theories or accurate news. However, memetics does help when it comes to explaining the spread of untrue, bizarre, and even harmful ideas. One such is alien abduction.

One key to alien abduction experiences is the phenomenon of sleep paralysis. During dreaming sleep most of our muscles are paralysed so that we do not act out our dreams. By the time we wake up this paralysis has usually worn off and we know nothing about it (unless experimenters intervene). However, the mechanism that keeps waking and dreaming apart can sometimes fail, especially in people who work shifts or have very disrupted sleep. Sometimes people wake up, can look around and think clearly, and yet are quite unable to move. Common sensations accompanying the paralysis include buzzing and humming noises, vibrations of the body or bed, a powerful sense that there is somebody or something in the room with you, and strange lights floating about. Since sexual arousal in dreams is common this can also persist into the paralysis. Sometimes, people feel as though they are being touched or pulled or even lifted from their bodies. If you recognise the symptoms, and can keep your cool, the best response is to relax and wait; the paralysis wears off within a minute or two. If you try to struggle you only make matters worse.

If you have no idea what is going on, the experience can be terrifying, and a natural reaction is to blame someone or something, or to seek an explanation. In previous periods of history and different cultures, various 'explanations' have been available. The incubus and succubus of medieval times were evil spirits sent to tempt the wicked into sexual activity. Until the early twentieth century, people in the south of England blamed witches for what they called 'hag-riding', and even today there are people in Newfoundland who have been visited by the 'Old Hag' who comes at night to sit on their chests and stop them breathing. Kanashibari in Japan, Kokma in St Lucia, and the Popobawa in Zanzibar are just some of the other current sleep paralysis myths. These myths are all successful memes.

Our culture is now full of stories of outer space, spaceships, UFOs, and sinister aliens. If you suffer from sleep paralysis and do not know what it is, your mind supplies the most readily available 'answer'. Once you start to think about aliens in a terrified and paralysed state, the aliens will seem all the more real. No wonder people think they have been abducted.

This explanation is supported by evidence that abductees suffer sleep disturbances, including sleep paralysis, more often than control groups (Spanos *et al.* 1993). I would expect, although it has not been tested, that people who understand the psychological basis of sleep paralysis are unlikely to have abduction experiences, because they already have a better explanation for their experience.

Some people have only faint memories of disturbing experiences and are left wondering what happened. If they come across a hypnotist who specialises in 'recovering memories' of alien abduction then they are encouraged to relive the experience again and again until the story they tell comes to be indistinguishable from a real memory and is full of details of the aliens and their craft.

But this is not the whole story. The myth of abduction is a successful set of memes for other reasons too. For one thing, it is extremely hard to test, which has protected it from being demolished. The aliens, after all, are so clever that they can slip in through your ceiling without disturbing the plaster, whisk you away, do their wicked experiments and put you back, all without anyone else seeing you, or them. They are also adept at suppressing your memories of the abduction, and you may be left only with a fragment of uncertain memory and a tiny unexplained scar on your leg or nose. It may take an experienced hypnotist (who has plenty of experience with other abductees and knows just what questions to ask) to bring your full 'memory' back. The aliens are rarely captured on radar or successfully photographed because they have such advanced technology. And if you want to know why no government has evidence of alien landings, well you know the answer – there's a conspiracy. Governments do, of course, have whole spaceships and even frozen alien bodies in store but they employ lots of people to ensure the evidence is concealed and the public is never allowed to know. If you wonder why none of these employees has ever let on, this just proves how powerful the conspiracy is. Interestingly, evidence that might appear to be *against* the myth, such as demonstrating that a claimed implant is really a filling, has almost no effect. Strong believers quite rightly point out that one piece of negative evidence cannot disprove their beliefs, and non-believers never thought it was an implant in the first place.

The abduction memeplex has been enormously successful and we can now see why. First it serves a genuine function. That is, it provides an explanation for a scary experience. I suspect that if my student had known about sleep paralysis before he had his first abduction experience, it never would have turned into an abduction experience. Second, the idea appeals in modern American (and to a lesser extent European) culture. Humans, like many of their primate relatives, have evolved to defer to high-status males and to be afraid of them. God thrives on this natural tendency but so too do the more modern powerful beings, who use the trappings of our scientific world and prey on genuine fears of technology. Third, the idea is promoted by television companies who have viewers eager to watch sensational programmes, and participants eager to tell their amazing, unique, fantastic, first-hand, real-life stories and to feel very special (and possibly even get rich) by doing so. And finally, the idea is more or less irrefutable, and protected by a more or less plausible conspiracy theory.

Just how good the protection is will determine how long the memeplex lasts. Like a virus, it will spread to infect as much of the susceptible population as it can reach and then, like a virus, stop spreading. Because its only genuine function depends on people being ignorant about sleep paralysis, the dissemination of scientific understanding of sleep may undermine it. Also, as many people ask for concrete evidence and none is forthcoming, the claims may eventually wear thin. Since this kind of television feeds on novelty and surprise, the producers will not keep asking abductees to come and tell their amazing stories forever. This particular memeplex, although successful, has a limited life. Others look a bit more secure.

Death and the truth trick

Research shows that people of all ages and backgrounds have somewhat similar experiences when they come close to death and are revived (Blackmore 1993). Although most experience nothing, those who do report experiences tend to describe going down a dark tunnel towards a bright light, leaving the body and viewing their body from above, travelling to a beautiful place where beings of light await them, occasionally experiencing scenes from their life replayed all at once, and finally having to make a difficult decision to return to ordinary life. Normally, the experience is happy and peaceful, although occasionally it can be hellish. Above all, it feels absolutely real – 'realer than real'. I have had this experience myself (although I was not close to death) and it was vivid, beautiful, absolutely realistic, and had a dramatic effect on my life. Reports from as long as two thousand years ago and from many different cultures suggest that the basic experience is common to humans everywhere and can affect them profoundly.

The core features of this complex experience can be understood in terms of what is going on in a brain under stress. For example, the surprisingly positive emotions are probably due to the brain releasing endorphins (morphine-like chemicals) under stress. Fear and stress also result in widespread random firing of neurons throughout the brain which, depending on where it occurs, has different effects. For example, stimulation of the temporal lobes (which can be done experimentally) can induce floating and flying sensations, memory flashbacks, and feelings of religious significance. Perhaps most interesting is the origin of the tunnel. Cells throughout the visual system are organised so that many are devoted to the centre of the visual field and far fewer to the periphery. So when all the cells fire randomly, the effect is like a bright light in the middle fading out towards the edges, or bands and lines in spirals and rings. This may be the origin of the near-death tunnel as well as the tunnels that are common in Shamanistic drawings and certain drug experiences.

Some near-death experiencers are happy to find rational explanations, but many reject them. They know they saw Jesus. He was as real as real can be. They know that they have a spirit that left their body, travelled down the tunnel and went to heaven. And they know their experience is proof of life after death.

What is interesting from the memetic point of view is that Christians typically see Jesus while Hindus meet Hindu deities (Osis and Haraldsson

1977). Some people encounter 'beings' of no particular religion, but there is no recorded case in which a religious person has met a deity from a different religion. Some Christians even meet St. Peter at his pearly gates, while Hindus are more likely to be judged by Chitragupta who has their names written in his great book. Americans are likely to go along with the heavenly beings they meet while Indians are more likely to resist when they meet Yamraj, the King of the Dead, or his messengers, the Yamdoots, who have come to take them away. Americans are likely to meet their mothers but Indians rarely encounter female figures.

The experienced 'realness' of the visions leads many people who have them to reject any naturalistic explanation at all, and in the scientific literature the debate is dichotomised between those who are convinced that near-death experiences (NDEs) are evidence of life after death and those, like myself, who are not (Bailey and Yates 1996). In fact, the experience cannot be evidence of life after death because all the people who described the experiences were still alive. Nor can any naturalistic explanation, however full and satisfying, prove there is *no* life after death. So the argument is ultimately sterile. But from the memetic point of view this is not the issue. Instead, we should ask a different question – why are NDE memes so successful?

The answer is similar to that for abductions. The NDE story serves real functions. First, there are underlying brain states that predispose certain kinds of experience when people come close to death and which cry out for explanation. These are interpreted with the memes available to that person at the time, whether those memes come from television, science or their religious upbringing. The classic NDE story also serves another function in reducing fear of death and providing reassurance about the meaning and purpose of life. Fear of death is a far more potent motivator than fear of sleep paralysis, and the desire for life after death an excellent hook for NDE memes. Memes do not need to be true to be successful.

They do, however, *claim* to be true. Natural selection has generally equipped us to choose ideas that are true over those that are false. Our perceptual systems are designed to provide as accurate a model of the external world as possible. Our capacity to think and solve problems is designed to give true rather than false answers, so in general, true memes should thrive better than false ones. But this provides an opportunity for deception – for truth mimicry. First, false claims can sneak into memeplexes under the protection of true ones. We might call this the 'truth trick'. Second, memes can simply *claim* to be true – or even 'The Truth'. So, for example, UFO believers claim that the conspiracy is suppressing The Truth. NDErs claim to have seen The Truth with their own eyes. And believers in God and life after death know The Truth. This is aslightly different version of the 'truth trick', for it need have no element of validity at all.

Finally, NDE memes use the 'altruism trick'. People who come close to death and survive are often changed by the experience, becoming more caring of others and less concerned with themselves (Ring 1992). The limited evidence available suggests that this change is a function of simply facing up to death, not of having a near-death experience, but when NDErs behave altruistically this helps spread their NDE memes – 'I'm a nice person, I'm not so selfish now, believe me. I really did go to heaven'. Wanting to agree with this honestly nice person helps spread the memes. And if the NDE survivor really does help you, then you may take on the NDE memes as a way of returning the kindness. Thus, NDE memes spread, and among them is the idea that people who have had NDEs behave more altruistically.

Other forms of the altruism trick are nastier. The Christian version of NDEs depends heavily on the idea that only good people go to heaven. Having a beautiful NDE implies you are a good person and should be believed. This also means that people who have hellish NDEs are less likely to report them, and their memes will do less well (not to mention the fear and loneliness they must feel if they cannot talk about their experiences). Disbelievers in life after death and researchers who pursue brain-based explanations are treated as nasty people who, if only they were nicer, would come to The Truth – another tactic that gives heavenly NDE memes the edge. No one wants to share the beliefs of a nasty person.

The most successful NDE memeplex in North America today is a rather sickly Christian version. Experiencers describe heavenly scenes, a classic Jesus, judgements based on the most narrow interpretation of morality, and lessons to be learned in this schoolroom of life. Their books stay on the best-seller lists for months and some of them become rich. In Europe, other versions seem to survive the competition a little better but, so far, scientific explanations are faring badly.

If we have to set naturalistic explanations against heavenly ones, then a memetic viewpoint is far more compatible with the former. But memetics cannot settle this impossible issue one way or the other. What it can do is explain why powerful myths spread through whole cultures and provide a shape for some of the most profound experiences of people's lives. These strange experiences are, like all our experiences, dependent on a brain state that has been shaped by both genes and memes. I suggest we will come to understand them better when we stop trying to draw a line between 'real' and 'unreal' experiences, and ask how particular experiences are created by natural and memetic selection.

From alien abductions and near-death experiences we can glimpse a general formula for certain kinds of successful memeplex. Take a highly emotional naturally occurring human experience with no satisfactory explanation, provide a myth that appears to explain it, and include a powerful being or unseen force that cannot easily be tested. As optional extras include other functions such as social coercion (the Old Hag gets you when you do wrong), reduction of fear (you'll live forever in heaven), use the altruism trick (good people have this experience or believe this myth) or the truth trick (this explanation is The Truth).

Until recently, no one designed such memeplexes on purpose. They were designed by memetic selection. We may imagine that hundreds of thousands of myths and stories have been invented over the millennia and passed on by thousands and millions of people. The few that survived were the ones that had all the good tricks to aid their recall and propagation. Modern culture is the legacy of thousands of years of memetic evolution.

Divination and fortune-telling

From magic crystals and Tarot cards to the healing power of aromatherapy and homeopathy, memeplexes spread using the tricks I have described, and some of their carriers get rich at other people's expense. Take Tarot cards, for example. Imagine you go to a Tarot reader and have the unnerving experience that she seems to know all about your life and personality and can give you advice over a problem that is troubling you. She seems to understand you uncannily well and gives details you think she could not possibly have known. Perhaps she says something like this (as you read this, try to imagine it is being said directly to you by a sincere looking, sympathetic woman who gives every sign of caring about you and your troubles, and looks deeply into your eyes between glances at the cards in front of her).

> You have a need for other people to like and admire you, and yet you tend to be critical of yourself. Disciplined and self-controlled on the outside, you tend to be worrisome and insecure on the inside. At times you have serious doubts as to whether you have made the right decision.
>
> I see from this card that you love animals. You have a cat, and the cards tell me you went to France last year. I know you are worried about the pain

> in your back but the orientation of this card shows that it will improve soon. I can see you playing as a child – now you may not know this yourself, but if you look carefully you will find you have a scar on your left knee.

Evidence suggests that Tarot readers succeed by (mostly unconsciously) using the perfectly normal skills of responding to feedback, reading subtle body language, and using the 'Barnum Effect' – that is, using statements that almost everyone will judge as true of themselves but not of others. I took the first three sentences from the classic Barnum Personality Reading (Forer 1949). Other Barnum statements include positive ones (few people would admit to not being kind-hearted), double-headed statements (one half is bound to be true of you) and ambiguous ones (read in what you like). Correct names and dates can be homed in on by trial and error in the certainty that clients will forget all the wrong tries, and will remember questions as though they were statements of fact. The little details I gave were ones I had heard so often that I included them in a survey of over 6000 readers of a British newspaper (Blackmore 1997). Among the results were that 29 per cent owned a cat, 27 per cent had been to France in the past year, 30 per cent were suffering pain in their back (and that's not including all the ones who might have had back pain in the past) and 34 per cent had a scar on their left knee (note the importance of scars in abductions too). You do not need to get every statement right to make a good impression.

So clients go away impressed and the Tarot reader becomes even more convinced of her powers, but that is not all. In the process, the clients pick up lots of Tarot card memes. The reader has special powers that you do not have. Tarot cards hold ancient mysteries that cannot be tapped by unbelievers. When you shuffle them they magically tune in with the rhythms of the universe and unfold your secret destiny. They will reveal the good in you and put you in touch with your higher nature – and so on.

These memes are successful because they seem to explain the client's experience and include all the right tricks. The fear they prey on is the fear of uncertainty and of making the wrong decisions in a horribly complex world. People typically go to psychics when they are at their lowest and want guidance. This means they are all the more likely to fall for claims of higher powers or of special insight. The 'illusion of control' also works in favour of these memes. Stress is reduced when control over a situation is increased – and if real control is not possible, an illusion of control will do (Langer 1975). Many experiments have shown the power of this illusion, and believers in the paranormal are more prone to it than disbelievers (Blackmore and Troscianko 1985). Similar arguments apply to the memeplexes associated with clairvoyance, palmistry, Feng Shui, divination with pendulums, and dowsing with twigs. Literally thousands of experiments have demonstrated without a shadow of a doubt that the claims of astrology are false (Dean *et al.* 1996) and yet one-quarter of American adults believe in the basic tenets of astrology and 10 per cent read an astrology column regularly (Gallup and Newport 1991). I think these disturbing facts are better explained in terms of the power of the memes to replicate themselves than by writing off so many people as simply stupid, ignorant or gullible.

Note the powerful use of the altruism trick in some New Age phenomena. Crystals imbued with special powers are created to help you; the ancient Egyptian food supplement will improve your life and fill you with natural vitality; a consultation with the colour therapist will harmonise your energies with the universe. The psychic is a spiritual person who is there only to help you (and does not really want to charge a fee). In fact, these methods of divination are just ways of appearing to see the future or read a person's mind, but they are routinely associated with goodness, love, compassion, and spirituality. We rarely ask the obvious question – what is 'spiritual' about astrology or a crystal ball? There is no obvious answer and yet these methods trade on that association. Bookshops categorise them all as 'Mind, Body and Spirit'. This is not good news for true compassion or spirituality. It is very good news for the moneymaking memes of the New Age.

I have deliberately chosen to tackle what some people might consider to be the most trivial memeplexes first. They may be trivial but they exert phenomenal power in modern society and are responsible for the movements of vast amounts of money. They shape the way we think about ourselves and, perhaps most importantly, they cause many people to believe things that are demonstrably false. Anything that can do all this deserves to be understood. The stakes are even greater when it comes to alternative medicine and the sale of ineffective therapies.

The sale of health

One survey estimated that every year Americans make 425 million visits to providers of unconventional therapy, spending over 13 billion dollars, and that 50 per cent of Americans use such therapies (Eisenberg *et al.* 1993). When alternative or complementary medicine is more narrowly defined, lower estimates (as low as 10 per cent) have been obtained, and it is claimed that the boom may now be over in Britain (Ernst 1998). Nevertheless, big money is at stake.

Some therapies may be effective in appropriate circumstances, such as relaxation, hypnosis, aromatherapy (massage with scented oils), and some kinds of herbal medicine. Others may work but not for the reasons usually given. For example, acupuncture works as an analgesic but the effect is now explained in terms of endorphins (the brain's own morphine-like chemicals) rather than the traditional Chinese theory of *ch'i* energy (Ulett 1992; Ulett *et al.* 1998). Chiropractic includes effective manipulations, although its traditional theory is false and it can sometimes be dangerous, and many other therapies use mixtures of the effective and ineffective. However, there are many therapies that are widely used and known to be completely useless or even harmful to health (Barrett and Jarvis 1993).

From a memetic point of view we need not ask why people are so stupid as to pay good money for demonstrably useless treatments, nor how intelligent people can be fooled by charlatans so easily, nor even how supposedly caring therapists can be so wicked as to promote false beliefs in vulnerable patients. Instead, we should look at what meme tricks these therapies are using. Then we can understand why they spread so quickly and get such a powerful grip on our society, when far more effective therapies do not. We do not even need to ask precisely which treatments work and which do not (though we should certainly do so when we are ill!). The validity of therapeutic claims is only one criterion for the success of memes, and there are many others. Once we start thinking this way the familiar signs are easy to see.

Alternative medicine preys on fear; fear of pain, fear of disease, and fear of death. It uses a natural human experience that (for most people) has no satisfactory explanation; that is, the experience of going to a therapist and feeling better. There is no doubt that people do generally feel better after a visit to an acupuncturist, herbalist, chiropractor, or homeopath. They have usually invested quite a lot of money in the visit or in the 'treatments' prescribed, and this is particularly effective in a country like England where conventional medicine is free on the National Health Service. 'Cognitive dissonance' theory explains why this is important – anyone who pays fifty pounds for a treatment that does not work will suffer the dissonance of concluding that they must be daft or have wasted good money – so an obvious way to reduce the dissonance is to convince yourself that you feel better (and note that the bigger the fee the better you must feel). The 'illusion of control' reduces stress, and hence some symptoms, because at least you are doing something about your health. Social pressures kick in when the therapist asks whether last week's treatment worked and you feel an obligation to say yes, or at least something encouraging. And once you say yes, the desire for consistency inclines you to convince yourself. The placebo effect is notoriously powerful and is increased when the therapist appears authoritative, and uses powerful sounding techniques, and impressive, if incomprehensible, explanations.

The explanations use a mixture of scientific-sounding terms and mysterious ones. Powerful beings and unseen forces are liberally invoked, including God and the spirits who act through the hands of the spiritual healer. The most commonly used word in alternative medicine is probably 'energy' – but the energy cannot be seen or tested. The *ch'i* of acupuncture and the 'innate intelligence' of chiropractic are so subtle that they cannot be investigated by any technique currently known to science, which neatly protects the memes from disproof. Finally, the altruism trick is liberally used, as when 'the power of love' is invoked. Alternative therapists are often genuinely caring people who really do want to help, and believe they are helping. Their patients tell them they feel better and so the therapists naturally (if falsely) conclude that their healing theory was right. Otherwise they may simply *appear* to be genuinely caring. Either way, the patient is more likely to take on their memes – false memes as well as true ones. All this amounts to a powerful formula for persistent and moneymaking memeplexes. No wonder there are so many of them around.

CHAPTER 15

Religions as memeplexes

Like it or not, we are surrounded by religions. The 'Great Faiths' of the world have lasted thousands of years and affect our calendars and holidays, our education and upbringing, our beliefs and our morality. All over the world people spend vast amounts of time and money worshipping their gods and building glorious monuments in which to do it. We cannot get away from religions, but using memetics we can understand how and why they have such power.

All the great religions of the world began as small-scale cults, usually with a charismatic leader, and over the years a few of them spread to take in billions of people all across the planet. Imagine just how many small cults there must have been in the history of the world. The question is why did these few survive to become the great faiths, while the vast majority simply died out with the death of their leader or the dispersal of their few adherents?

Dawkins was the first to give memetic answers (Dawkins 1986, 1993, 1996*b*), although his ideas on religion have frequently been criticised (Bowker 1995; Gatherer 1998). He took Roman Catholicism as an example. The memes of Catholicism include the idea of an omnipotent and omniscient God, the belief that Jesus Christ was the son of God, born of the virgin Mary, risen from the dead after his crucifixion and now (and for ever) able to hear our prayers. In addition, Catholics believe that their priests can absolve them from sins after confession, the Pope literally speaks the word of God, and when priests administer the mass, the bread and wine literally change into the flesh and blood of Christ.

To anyone uninfected with any Christian memes these ideas must seem bizarre in the extreme. How can an invisible God be both omnipotent and omniscient? Why should we believe a two-thousand-year-old story that a virgin gave birth? What could it possibly mean to say that the wine 'literally' becomes the blood of Christ? How could someone have died for our sins when we were not even born? How could he rise from the dead, and where is he now? How could a prayer, said silently to yourself, really work?

There are many claims for the efficacy of prayer in healing the sick, and even a little experimental evidence (Benor 1994; Dossey 1993), but few of the experiments have controlled adequately for placebo effects, expectation, and spontaneous recovery, and some have shown that people with the strongest religious faith were *less* likely to recover from acute illness (King *et al.* 1994). Against the claims are hundreds of years of people praying for the health of their royal families or heads of state with no apparent effect, and the inability of modern-day religious healers to make any obvious difference in hospitals. Then there are all those countless wars in which both sides routinely pray for God to help their side and kill the enemy. Yet millions of people all over the world profess themselves Catholics and pray to Jesus, his mother Mary, and God the Father. They spend vast amounts of their valuable time and money supporting and spreading the faith to others, and the Catholic Church is among the richest institutions in the world. Dawkins (1993) explains how religious memes, even if they are not true, can be successful.

The Catholic God is watching at all times and will punish people who disobey His commandments with most terrible punishments – burning forever in hell, for example. These threats cannot easily be tested because God and hell are invisible, and the fear is inculcated from early childhood. A friend of mine showed me a book he once treasured as a child. It had pictures of a little good boy and a little bad boy. You could open up the flaps of their blazers and inside the good boy find a white and shining heart, while the bad boy had a black spot for every sin he had committed. Imagine the power of that image when you cannot see inside your own body and must only imagine the little black spots piling up and piling up – when you talk in class or cheat in a test, when you take your sister's toy or steal a chocolate biscuit, when you think a bad thought, or doubt God's truth and goodness . . . every one a black spot.

Having raised the fear, Catholicism reduces it again. If you turn to Christ you will be forgiven. If you honestly repent of your sins, bring up your children as Catholics, and go regularly to mass, then, even though you are unworthy and sinful, God will forgive you. God's love is always available but at a price, and that price is often overlooked completely because it is paid so willingly. It is the price of investing massive amounts of time, energy and money in your religion – in other words, working for the memes. As Dawkins pointed out, Catholics work hard to spread their Catholicism.

I previously described several meme tricks that New Age memeplexes use. All these can be found in religions too. First, like alien abduction and near-death experience memes, religions serve a real function. They supply answers to all sorts of age-old human questions such as: Where do we come from? Why are we here? Where do we go when we die? Why is the world full of suffering? The religious answers may be false but at least they are answers. Religious commitment may give people a sense of belonging, and has been shown to improve social integration in the elderly (Johnson 1995). Religions may also incorporate useful rules for living, such as the dietary laws of Judaism or rules about cleanliness and hygiene which may once have protected people from disease. These useful functions help carry other memes along.

The truth trick is liberally used. In many religions, God and Truth are virtually synonymous. Rejecting the faith means turning away from Truth; converting others means giving them the gift of the true faith. This may seem odd when so many religious claims are clearly false, but there are many reasons why it works. For example, people who have a profound experience in a religious context are inclined to take on the memes of that religion; people who like or admire someone may believe their truth claims without question. At the extreme, people will even tell lies for God and manage to convince themselves and others that they do so in the name of truth – as when 'Creation Scientists' proclaim 'The Truth' that the earth is only six thousand years old, and back it up with denials of the fossil record, or claims that the speed of light has slowed since the creation so as to give the illusion of a vast universe and an ancient planet (Plimer 1994).

Beauty inspires the faithful and brings them closer to God. Some of the most beautiful buildings in the world have been constructed in the name of Buddha, Jesus Christ, or Mohammed. Then there are the beautiful statues and alluring stories in Hinduism; stained glass, inspiring paintings, and illustrated manuscripts; uplifting music sung by tremulous choir boys and vast choirs, or played on great organs. Deep emotions are inspired to the point of religious ecstasy or rapture which then cries out for – and receives – an explanation. The ecstasy is real enough, but from the memes' point of view, beauty is another trick to help them reproduce.

The altruism trick permeates religious teachings. Many believers are truly good people. In the name of their faith they help their neighbours, give money to the poor, and try to live honest and moral lives. If they are successful then generally people come to like and admire them and so are more inclined to imitate them. In this way not only does good and honest behaviour spread, but the religious memes that were linked to that behaviour spread too. Alongside this comes merely the semblance of good behaviour. Hypocrisy can flourish when goodness is defined not only as kind and altruistic behaviour, but as sticking to the rules and obligations of the faith. Much of the money donated to churches, temples, or synagogues is not used for the poor or needy, but to perpetuate the religion's memes by erecting beautiful buildings or paying for clergy. Activities that spread memes are also defined as 'good' even though their benefit is questionable, such as saying prayers at specified times, saying grace at every meal, and keeping one day of the week as a day of worship. In this way huge chunks of every believer's time are willingly devoted to maintaining and spreading the faith.

Many people think of Mother Teresa as a saint. Indeed, she may soon be officially canonized by the Catholic Church. She is many people's idea of the truly selfless and altruistic heroine. But what did she actually do? Some of the inhabitants of Calcutta accuse her of diverting attention from the real needs of the city's poor, of giving Calcutta a bad name and of helping only those who were prepared to take on Catholic teachings. Certainly, she was fiercely anti-abortion and anti-birth-control. Many of the people she helped were young women with no access to contraceptives, little ability to avoid being raped, and almost no access to health care if they became pregnant. Yet she steadfastly maintained her Catholic opposition to the one thing that would have helped them most of all – control over their own reproductive lives. Whatever we may think about how much she really helped the starving people of Calcutta there is no doubt that her behaviour effectively spread Catholic memes by using the altruism trick.

Even evil and cruelty can be redefined as good. The Koran states that it is good to give a hundred lashes to an adulteress and to have no pity on her. You might think that Muslim women can avoid this by not committing adultery, but Warraq (1995) explains in unpleasant detail what life can be like in countries that adhere strictly to Islamic law. Women may be powerless to resist sexual abuse, and afterwards must take the punishment while the men who abused them get off free. Since women are objects of disgust, a man is supposed not to touch a woman he does not have rights over. Women are routinely locked away and, if they are allowed out, must walk behind the man and be suitably covered – which in many countries means being covered head to toe in a smothering garment with just a tiny grille to look out of. Obeying such rules to the letter makes a Muslim 'good', regardless of the misery it creates.

Returning to more honest uses of goodness and altruism, Allison's (1992) theory of 'beneficent norms' applies especially well to religions. One of his general rules is 'Be good to your close cultural relatives'; the memetic equivalent of kin selection. But how do you know who they are? This rule tracks biological kinship in cultures with predominantly vertical transmission, since in these cultures you acquire most of your memes from biological relatives, but with horizontal transmission other means of recognition are needed. One is 'Be good to those who act like you'. It works like this. If you see someone else who acts the same way as you do, it is likely that you both have cultural ancestors in common. If you now help him you make it more likely that he will be successful, and hence that he will pass on his memes, including the rule 'Be good to those who act like you'. Allison calls this a 'marker scheme'. He gives the examples of wearing a turban or abstaining from certain foods, but we might add supporting Manchester United or listening to hip-hop, as well as genuflecting or wearing a little portrait of your guru round your neck. He adds that markers that are costly or difficult to learn can deter exploitation by outsiders. Apart from languages, a good example is religious rituals. Many of these require years to learn and others, such as ritual circumcision, are certainly costly for an adult.

The result of this kind of altruism is that people are kind and generous to the in-group and not to outsiders. This boosts the well-being of the group's members and hence makes them more likely to be imitated, and so pass on the faith. This is exactly what we see in many of the world's greatest religions. Although the instruction to 'love thy neighbour as thyself' is commonly taken to mean 'love everyone', in the tribal context in which it was first written it may have been meant more literally – in other words love your own tribe, and your own family, but not everybody else (Hartung 1995). Even the admonition not to kill may originally have applied only to the in-group. Hartung points out that the rabbis of the Talmud used to hold an Israelite guilty of murder if he intentionally killed another Israelite, but killing other people did not count.

Some religions positively encourage murder and war against people of other faiths. Islam has fatwas and jihads to justify killing unbelievers, and especially those who harm or renounce the faith. In February 1989, the Ayatollah Khomeini delivered his famous fatwa on the author Salman Rushdie. This is a direct call to all Muslims to murder Rushdie for daring to blaspheme against the holy Koran in his book *Satanic Verses.* When the punishment for renouncing or criticising a religion is so severe, the memes are very ably protected.

Hindus, Muslims, and Christians alike have gone to war again and again in the name of God. When a few hundred Spaniards murdered thousands of Incas, leading to the destruction of an entire civilisation, they did it for the glory of God and the holy Catholic Faith. In a subtler way religious missionaries are still destroying ancient cultures even today. People have been tortured, burned alive, and shot because they believed the wrong thing. Religions teach that God wants you to spread his *True* understanding to all the world and it is therefore *good* to maim, rape, pillage, steal, and murder.

We saw how the conspiracy theory protects UFO memes; similar mechanisms protect religious memes. As Dawkins (1993) points out, good Catholics have faith; they do not need proof. Indeed, it is a measure of how spiritual and religious you are that you have faith enough to believe in completely impossible things without asking questions, such as that the wine is *really* turned into blood. This assertion cannot be tested because the liquid in the cup still tastes, looks and smells like wine – you must just have faith that it is *really* Christ's blood. If you are tempted by doubt, you must resist. Not only is God invisible but he 'moves in mysterious ways'. The mystery is part of the whole package and to be admired in its own right. This untestability protects the memes from rejection.

Religious memes are stored, and thus given improved longevity, in the great religious texts. The theologian Hugh Pyper (1998) describes the Bible as one of the most successful texts ever produced. 'If "survival of the fittest" has any validity as a slogan, then the Bible seems a fair candidate for the accolade of the fittest of texts' (p. 70). It has been translated into over two thousand languages, exists in many different versions within some of those languages, and even in a country like Japan, where only one or two per cent of the population are Christians, more than a quarter of all households possesses a copy. Pyper argues that Western culture is the Bible's way of making more Bibles. And why is it so successful? Because it alters its environment in a way that increases the chances of its being copied. It does this, for example, by including within itself many instructions to pass it on, and by describing itself as indispensable to the people who read it. It is extremely adaptable, and since much of its content is self-contradictory it can be used to justify more or less any action or moral stance.

When we look at religions from a meme's eye view we can understand why they have been so successful. These religious memes did not set out with an intention to succeed. They were just behaviours, ideas and stories that were copied from one person to another in the long history of human attempts to understand the world. They were successful because they happened to come together into mutually supportive gangs that included all the right tricks to keep them safely stored in millions of brains, books and buildings, and repeatedly passed on to more. They evoked strong emotions and strange experiences. They provided myths to answer real questions and the myths were protected by untestability, threats, and promises. They created and then reduced fear to create compliance, and they used the beauty, truth and altruism tricks to help their spread. That is why they are still with us, and why millions of people's behaviour is routinely controlled by ideas that are either false or completely untestable.

....

No one designed these great faiths with all their clever tricks. Rather, they evolved gradually by memetic selection. But nowadays people deliberately use memetic tricks to spread religions and make money. Their techniques of memetic engineering are derived from long experience and research, and are similar to those used in propaganda and marketing; with radio, television and the Internet, their memes can spread far further and faster than ever before. Billy Graham's style of tele-evangelism is a good example. He starts by evoking fear, reminding people of all the terrible things happening in the world and of their own impotence and mortality. He presents science as having no answers and as a cause of the world's ills, and then persuades people to surrender to the all-powerful God who is their only hope of salvation. The experience of surrender raises powerful emotions and people turn to God in huge numbers.

Other evangelists use healing to spread the Word. We have seen how perfectly normal psychological processes can make people feel better, even when they are not actually cured, and this is a powerful incentive to take on the God memes that often accompany the healing. The trip to Lourdes is expensive and difficult. Expectations are high. Spiritualist healers are kind and plausible, and really do seem to care about your troubles.

Some use fake healing. In the 1980s, Peter Popoff and his wife Elizabeth brought millions of Americans to God, and millions of dollars to the Popoffs, through their healing missions. Their vast audiences sang and prayed, and watched seriously ill people stagger onto the stage, raising powerful emotions as the Popoffs appealed for donations. As Peter correctly diagnosed illnesses and announced the sufferers cured, people forgot that an hour before Elizabeth had wandered through the audience collecting prayer cards on which people wrote their names, addresses, ailments and other crucial facts. She took these to the computer database backstage and beamed the information to a receiver behind Peter's left ear (Stein 1996).

Miracles of all kinds have been used to convert unbelievers. Jesus walked on water and brought a dead man back to life, nineteenth-century spiritualist mediums created spirit forms made of 'ectoplasm', and the advanced practitioners of transcendental meditation claim to levitate.

Some people effectively combine special powers with the altruism trick, such as England's much-loved grandmotherly medium Doris Stokes who packed her audiences with clients whom she already knew, and fooled millions (I. Wilson 1987). Many of those clients were recently bereaved wives, husbands or parents who gained comfort from Stokes's messages but who might have coped better with their grief if they had been helped to accept the reality of death.

I do not mean to imply, from all I have said, that there are no true ideas anywhere in any religion. The memetic mechanisms I have described would allow religions to flourish that were based on complete falsehoods and nothing else, but there may be true ideas embedded in them as well. Just as some alternative therapies thrive by including a few treatments that work, so religions may include valid insights as well as misleading myths.

At the heart of many religions lie the mystical traditions, like that found in the fourteenth-century *Cloud of Unknowing* or the teachings of Julian of Norwich in Christianity; the Sufi teachings of Islam; or the stories of enlightenment in Buddhism. These traditions emphasise direct spiritual experience which is often ineffable and therefore hard to pass on. In spontaneous mystical experiences people typically feel they have been given a glimpse of the world as it really is. They feel that self and other have become one, the entire universe is as it is, or that everything is oneness and light. These may indeed be valid insights (I believe they are), but on their own they are not very successful as memes, and rapidly get overtaken by all the more powerful religious ideas I have described above.

Buddhism provides a good example. If the stories are to be believed, the Buddha sat under a tree, with a fervent desire to understand, until finally he became enlightened. He then taught what he had seen, that everything is empty of self-nature, that life is unsatisfactory, that suffering comes about through craving or attachment, and that the cessation of craving leads to freedom from suffering. He laid down an ethical code of behaviour and taught his disciples to work out their own salvation with diligence, by calming the mind and practising attention in every moment. None of this is very comforting. Basically, it means you are on your own in a fundamentally unsatisfactory world with no one to help you. If you look to anything at all to try to make it better then you are caught up in craving and hence suffering. Enlightenment is not something to be attained; it is simply the giving up of – well everything really. As one of my students put it 'I couldn't bear not to want chocolate. I couldn't even imagine not craving chocolate, let alone not craving anything.'

So what happens to difficult ideas like these? Perhaps surprisingly they can and do survive, often by being passed in an unbroken chain from inspiring and enlightened teachers to hard-working pupils. Zen Buddhism sticks quite closely to the simplest teachings and includes no deities or hidden powers; no altruism nor beauty tricks. One is told to find out the truth for oneself and trained simply to sit and watch the mind until it becomes clear. These difficult ideas have survived almost dying out in the East and are now spreading widely in the West (Batchelor 1994). However, other forms of Buddhism are much more popular all over the world, such as Tibetan Buddhism, with its numerous powerful deities, beautiful buildings and paintings, stories of marvellous deeds, reciting of sutras, chants, and liturgies. Whether or not there are true insights at the heart of any religion, the fact is that clever memes will tend to beat them in the battle for replication.

We can now see how and why religions have the power and persistence they do. I want now to consider two further questions. First, have they played any part in meme–gene coevolution? And second, how are religions changing now that memes are being spread by modern technology?

The coevolution of religions and genes

The coevolutionary question is this. Have the religious memes that thrived in the past had any effect on which genes were successful? If so, this would be another example of memetic driving. I shall speculate here and hope that some of the questions I raise may be answered by future research.

We know little of the earliest religions. There is evidence of burial of the dead from the Neanderthals who lived from 130 000 to 40 000 years ago, but it is likely that they were not our ancestors. About 50 000 years ago came what is sometimes called the 'Great Leap Forward', characterised by improvements in toolmaking, the beginnings of art, and the creation of jewellery which was sometimes buried with the dead. We can only guess at religious beliefs but burial rites at least suggest some idea of an afterlife. Modern hunter-gatherer societies have varied religious beliefs, including ancestor worship, special powers attributed to the priest or shaman, and belief in an afterlife. So we might guess that early human religions were something like this.

Early humans lived in bands or tribal societies and only gradually did more complex stratified societies evolve. In chiefdoms or states there is enough division of labour for some people to be completely freed from food production; these are typically rulers of various kinds, together with soldiers and priests. Diamond (1997) argues that the function of ideologies and religions in chiefdoms is to justify the redistribution of wealth, the authority of the rulers, and warfare. Chiefs typically take enormous amounts of wealth from working people and use some of it to build grand temples or public works as visible signs of their power. The people may accept their wealth being taken from them, as they accept taxation in modern societies, if they obtain benefits in return. These benefits may include the reduction of violence within the society, protection from enemies, or facilities for public use. Sometimes the ruler and priest are the same person, but in larger societies separate priests take on the religious functions. The priests promote and police the religious beliefs; the beliefs are then used to justify the conquest of other peoples from whom more goods and power can be stolen.

In memetic terms what this amounts to is that the religious memes are more likely to survive and replicate than competing memes are. For example, religions that required no priests, that took no taxes, or that built no impressive buildings, would have been at a disadvantage. This meant the proliferation of highly organised and stratified societies and of priests who taught and maintained the religion. Religious memes have therefore played an important role in the development of human societies.

The coevolutionary question is whether they have affected the genes along the way. E. O. Wilson (1978) treated religions as a challenge to his new science of sociobiology and speculated about the ways in which religious belief could provide a genetic advantage. For example, religions often include prohibitions against eating potentially contaminated foods, and against incest and other risky sexual activities, and encourage believers to have large and well-protected families. In these and other ways religious belief would benefit the genes of believers and so be expected to continue. The evolutionary psychologist Steven Pinker (1997) has argued that religious beliefs are by-products of the brain modules that were designed to do other things; spirits and gods are based on our concepts of animals and people; supernatural powers are inferred from natural powers; the idea of other worlds is based on dreams and trances. As he puts it: 'religious beliefs are notable for their lack of imagination (God is a jealous man; heaven and hell are places; souls are people who have sprouted wings)' (Pinker 1997, p. 557). These authors argue either that religions provide a genetic advantage, or that they are the by-product of things that once provided genetic advantage. They do not consider the possibility of memetic advantage, nor of memes driving genes.

There are several ways in which memes might have influenced genes. Priests attain power and status by predicting (or appearing to predict) weather, disease, or crop failures; by building or being associated with temples and other grand buildings; by wearing expensive and impressive clothes; and by claiming supernatural powers. In many cultures the priests or rulers are given divine status. We know that women prefer to mate with high-status men, and that these men leave more offspring, either by having more wives or by fathering children by women who are not their wives. Even in societies in which the priesthood is celibate and could not (or at least should not) pass on their genes, other people could acquire power by association. If this religious behaviour helped people acquire more mates, then any genes that inclined them to be more religious in the first place would also flourish. In this way *genes* for religious behaviour would increase because of religious *memes*.

The idea of 'genes for religious behaviour' is not at all implausible – all it means is genes that make people more inclined towards religious beliefs and behaviour. Brain development is under genetic control and it is known that some brains are more prone to religious belief and experience than others. For example, people with unstable temporal lobes are more likely to report mystical, psychic and religious experiences, and to believe in supernatural powers, than those with stable temporal lobes (Persinger 1983). Like many other psychological variables, religiosity is known to have a heritable component even today. For example, identical twins are more similar in religiosity than non-identical twins or siblings. In our past there may have been as much genetically controlled variation in religious behaviour as there is now, or even more. If so, two effects are possible. First, the memetic environment could have influenced whether genes for religious behaviour were positively selected or not (increasing or decreasing religious behaviour in general). Second, the religion of the time could have influenced the *kinds* of genes that survived (i.e. those that produced the kind of religious behaviour best suited to that religion). That would be memetic driving at work.

Group selection

There is another way in which religious memes might conceivably drive the genes: through group selection. The whole concept of group selection has had a troubled history and been beset by controversy. Earlier this century it was invoked to explain all kinds of behaviours that might conceivably benefit groups or societies, and biologists often appealed to 'group adaptations' or 'the good of the species' without any idea of possible mechanisms. Williams's classic book *Adaptation and Natural Selection* (1966) pointed out the errors: for example, that selfish individuals could always infiltrate altruistic groups and thrive at their expense. Also groups have a slow lifecycle compared with individuals, and individuals can often move between groups. This means that individual adaptations will almost always predominate over adaptations for the group. Therefore, we should not look to group selection as a force that can make individuals sacrifice their own genetic interests 'for the good of the group'.

Most biologists now consider that group selection is only a weak force in nature (Mark Ridley 1996). However, selection at the level of the group can sometimes occur. Dawkins's distinction between the replicator and the vehicle is helpful here. In most of biology the replicator (the thing that gets copied) is the gene, while the vehicle is the whole organism. Whole organisms; that is individual cats, donkeys, orchids or cockroaches, live or die, and in the process either pass on their genes or not. All the genes in that vehicle share the same fate. In this (the most common) case selection is taking place at the level of the organism.

In some cases, however, whole groups of organisms live or die, and so all the genes in the group are killed off at once. If this occurs then the group is the vehicle and we can say that selection is happening at the level of the group. This applies, for example, to whole species that go extinct, or to isolated populations of animals, such as those on small islands, in which some groups survive and some do not. In these cases there is no conflict between individual and group selection (as there was in the argument about altruistic behaviour) but selection has acted at the level of the group.

Ridley (1996) concludes that group selection works only if migration rates are implausibly low and group extinction rates implausibly high. Another way of putting it is that group selection is favoured by mechanisms that reduce the differences in biological fitness *within* the groups and increase differences *between* groups, thus concentrating selection at the group level (D. S. Wilson and Sober 1994).

Memes may provide just this kind of mechanism. Indeed Boyd and Richerson (1990) have used mathematical models to show that group selection is particularly likely to occur when behavioural variation is culturally acquired, and that it can even occur with large groups and substantial rates of migration. The important point is that memes can have precisely the effect of decreasing within-group differences and increasing between-group differences.

Let us take dietary habits as an example. Suppose that one group of people eat shellfish as a major part of their diet and develop ways of cooking mussels or clams and getting them out of their shells, while another group of people hold a taboo against eating shellfish. People within each group are more similar to each other, and different from people in the other group. Migration between groups is made difficult by long habits of taste, and the difficulty of learning how to prepare the food. In some environments the first group may do better because they get more protein, while in other environments the second group may do better because they survive a lethal disease from infected food. When disease strikes or famine threatens it is whole groups that live or die. Food taboos are an important part of many religions. Orthodox Jews do not eat shellfish or pork and avoid mixing meat with milk. Many Buddhists and Hindus are vegetarians because they do not want to kill animals. The beliefs that underpinned these taboos may have caused some groups of people to survive and others to go extinct; and both their genes and their memes would have gone with them.

Religions also dictate sexual practices, promote certain kinds of cooperative behaviour, and regulate aggression and violence. Although many people believe that primitive tribes live an idyllic and peaceful existence, this myth (like so many in anthropology) has been exploded. The anthropologist Napoleon Chagnon (1992) lived for many years with the Yanomamö, who live in the Brazilian rain forest by hunting and growing food in temporary gardens. He describes a violent life in which war between villages is common and murders are revenged with more murders. Similar stories come from many parts of the world. In New Guinea, a group of nomads called the Fayu live in small family groups who only rarely meet other families because of the revenge murders that ensue when they do. Gatherings, for example to exchange brides, are fraught with danger. In many tribal societies murder is a leading cause of death (Diamond 1997). Although many people in modern cities believe that they face ever increasing risks of being killed they are in fact far safer than they would have been in a band or tribal society. The organisation that comes with government and religion therefore decreases these kinds of violence. However, it also provides the justification for large-scale wars.

The history of warfare is largely a history of people killing each other for religious reasons. Religions give people a motive, other than genetic self-interest, for sacrificing their lives for others – something that does not happen in band and tribal societies. Young men may believe that it is good to die for God, heroic to be killed in a religious war, or that they will have a place reserved for them in heaven. A society in which brave young men are prepared to die for their beliefs is likely to win a war against a society in which they are more concerned about protecting themselves or avenging their family. Such a victory is a victory for the memes that created the difference in the first place, and for the genes of the survivors.

We can now see why group selection might be important in memetics. Religions are a good example of a mechanism that decreases within-group differences, while increasing between-group differences and rates of group extinction. In many religions conformity is encouraged, forbidden behaviours are punished, differences between believers and unbelievers are exaggerated, fear or hatred of people with other beliefs is nurtured, and migration to a different religion made difficult or impossible. Wars between religious groups are common and in our evolutionary history many groups have lived or died for their religion. All this makes it more likely that group selection has occurred. If there were genetic differences between the groups to start with, then the survival of some groups and extinction of others would have had effects on the gene pool. In this case we could say that the religious memes have driven the genes.

This is likely to be most interesting if, for example, there was some genetic reason why one group took up one religion while a different group took up a different religion. Let us imagine two neighbouring groups of early hominids in which, by chance, one group had more of a genetic tendency to want to bury their dead in elaborate ways. This is not at all far-fetched if you remember that digging and burying behaviour is under genetic control in many species, from worms and wasps to rabbits and dogs. This genetic propensity then encouraged these people to develop a religion based on ancestor worship and an afterlife – we can call them the 'Afterlifers'. Meanwhile, the other group developed a religion based on worshipping nature spirits – we can call them the 'Naturists'. The Afterlifers then developed a taste for war, believing their ancestors' spirits would aid them, and that they would individually go to heaven if they killed an enemy, whereas the Naturists just got on with their own interests. In consequence, the Afterlifers won more of the wars against the Naturists; their memes spread – and so did their genes. Genes for the original ritual burying behaviour were selected for by group selection driven by memes.

I am not suggesting that this precise series of events has actually happened, but that this general mechanism could have shaped human nature and given us our religious tendencies. The principle is a general one and could theoretically apply to all kinds of genetic predispositions, such as conformity, having religious experiences, enjoying ritual and worship, or believing in life after death. This process could even have acted to favour genes that would otherwise be detrimental to fitness, or to wipe out genes that would otherwise have been fitness-enhancing. So some aspects of human nature could have been determined not for the sake of the genes but for the sake of the memes. Our beliefs could have moulded the way genetic selection took place. If this has happened it means that human beings might now be naturally religious creatures because of our long memetic history.

Religions have held enormous power for millennia, but times are changing and religions with them. One obvious change is that vertical transmission is giving way to the faster horizontal transmission (p. 132). As people are increasingly exposed to new ideas from television, radio, newspapers and the Internet, they begin to make comparisons and ask difficult questions. So it is, sadly, not surprising to learn that Afghanistan's Taliban Islamic movement has forbidden televisions and radios, and has set about destroying any they find, and punishing their owners. Meanwhile, in countries with thriving communications, some of the tricks the old religions use may not work so well any more. When people can see films, go to art galleries, and listen to any music they like, the beauty trick is less effective. When we are subjected on television to the gruesome results of religious wars, the altruism trick wears thin. When Christian leaders argue over whether homosexuality is *really* a sin, the truth trick begins to weaken its grip.

In the past, religions that promoted large families were successful because they created more people to adopt the faith from their parents. Lynch (1996) has given many examples of religions, from the ancient Islam to the relatively new and thriving Mormonism, that spread by increasing the number of their offspring, but he does not clearly differentiate the effects of vertical and horizontal transmission. With modern horizontal transmission people are less bound by their parents' beliefs; as memes spread faster and faster the birth rate becomes less and less significant. We should therefore expect proselytic religions to do better in technologically advanced societies. We may expect new religions of this kind, and also that old faiths which can adapt their memes to changing times may survive while others will become extinct.

I doubt that human beings will ever be entirely free of religion. If the arguments above are right then religions have two very strong forces going for them. First, human minds and brains have been moulded to be especially receptive to religious ideas, and second, religious memes can use all the best meme tricks in the book to ensure their own survival and reproduction. This may explain the persistence of religion in scientifically literate societies and in societies in which political dogma has tried to erase all religious behaviour – and failed. Perhaps our brains and minds have been moulded to be naturally religious and it really is difficult to use logic and scientific evidence to change the way we think – difficult, but not impossible.

Science and religion

I have implied that science is, in some sense, superior to religion, and I want to defend that view. Science, like religion, is a mass of memeplexes. There are theories and hypotheses, methodologies and experimental paradigms, intellectual traditions and long-standing false dichotomies. Science is full of ideas that are human inventions, and have arbitrary conventions and historical quirks built into them. Science is not 'The Ultimate Truth' any more than any other memeplex. However, memetics can provide a context in which to see why science offers a better kind of truth than religion.

We are designed by natural selection to be truth-seeking creatures. Our perceptual systems have evolved to build adequate models of the world and predict accurately what will happen next. Our brains are designed to solve problems effectively and to make sound decisions. Of course, our perception is partial and our decision-making less than brilliant – but it is a lot better than useless. If we had no memes, that would be that; we would have the best understanding of the world that could be acquired in the circumstances. But we do have memes, and with memes come not only new ways of controlling and predicting the world, but meme tricks and free-loading memes, misleading memes and false memes.

Science is fundamentally a process; a set of methods for trying to distinguish true memes from false ones. At its heart lies the idea of building theories about the world and testing them, rather like perceptual systems do. Science is not perfect. Scientists occasionally cheat to gain power and influence, and their false results can survive for decades, misleading scores of future scientists. False theories thrive within science as well as within religion, and for many of the same reasons. Comforting ideas are more likely to last than scary ones; ideas that exalt human beings are more popular than those that do not. Evolutionary theory faced enormous opposition because it provided a view of humans that many humans do not like. The same will probably be true of memetics.

However, at the heart of science lies the method of demanding tests of any idea. Scientists must predict what will happen if a particular theory is valid and then find out if it is so. That is precisely what I have tried to do with the theory of memetics.

This is not what religions do. Religions build theories about the world and then prevent them from being tested. Religions provide nice, appealing and comforting ideas, and cloak them in a mask of 'truth, beauty, and goodness'. The theories can then thrive in spite of being untrue, ugly, or cruel.

In the end, there is no ultimate truth to be found and locked up forever, but there are more or less truthful theories and better or worse predictions. I do defend the idea that science, at its best, is more truthful than religion.

CHAPTER 16

Into the Internet

In our house we have four telephone lines, two fax machines, three television sets, four hi-fi systems, seven or eight radios, five computers and two modems. And there are only four of us. We also have many thousands of books and a few compact disks, audiotapes and videotapes. How did all this stuff come to exist and why?

If you have never asked yourself the question you might think the answer is obvious. All these things are great inventions, created by other people to make our lives better or more fun. But is this the right answer? Memetics provides an entirely different answer, one that is somewhat counter-intuitive.

I suggest that memetic selection created them. As soon as memes appeared they started evolving towards greater fidelity, fecundity, and longevity; in the process, they brought about the design of better and better meme-copying machinery. So the books, telephones, and fax machines were created by the memes for their own replication.

This may sound odd when we know that memes are just information being copied from one person to another. How can bits of information create radios and computers? But the same question could be asked of genes – how can bits of information stored in DNA create gnats and elephants? The answer is the same in both cases – because the information is a replicator that undergoes selection. This means the evolutionary algorithm runs, and the evolutionary algorithm produces design. The design of computers by memetic selection is, in this sense, no more mysterious than the design of forests by genetic selection. The consciousness of a designer is not the causal factor in either process. Design comes about entirely from the playing out of the evolutionary algorithm.

. . . .

We are used to the idea of animals and plants being designed by natural selection, but we must also think about the evolution of the replication machinery that makes natural selection possible – for both have evolved together. This is the analogy I wish to draw here. Memes do not yet have precise copying machinery as DNA has. They are still evolving their copying machines and this is what all the technology is for.

It is helpful to look back at what must have happened in the case of genes – the only other replicators we know much about (Maynard Smith and Szathmáry 1995). When the first ever replicator arose on this planet it was presumably not DNA but some simpler precursor, or even some completely different replicating chemical. Whatever it was, we can be sure that the cellular machinery for copying it did not exist. Natural selection in the very early days of life was not selecting between complex organisms like cats and dogs, or even different kinds of simple cell, but between little bits of protein or other chemicals. Any of these proteins that got copied more often or more accurately, or that lasted longer, would have survived at the expense of the rest. Gradually, from these beginnings, natural selection would produce not only more proteins but proteins that took part in the copying of other proteins. Eventually, there evolved the system of groups of replicators, replicating machinery, and vehicles that we see today. The system settled down so that all creatures on the planet use the same (or a very similar) replication system which produces extremely high-fidelity copying of long-lasting replicators.

I suggest that the same process is now going on with memes, except that it is still in its infancy. As Dawkins put it, the new replicator is 'still drifting clumsily about in its primeval soup' (Dawkins 1976, p. 192). That soup is the soup of human culture, human artefacts, and human-made copying systems. You and I are living during the stage at which the replication machinery for the new replicator is still evolving, and has not yet settled down to anything like a stable form. The replication machinery includes all those meme-copying devices that fill my home, from pens and books to computers and hi-fi.

Looking at it this way we can see all sorts of critical inventions of human culture as phases in the evolution of meme replication. I have already explained how treating language this way provides a new theory of the origins of language. I want now to go on from spoken language itself to the invention of writing, and then to modern information-processing technology. As before, we should expect the evolutionary process to involve increases in the fidelity, fecundity, and longevity of the replicators.

Writing

Writing is obviously a useful step for memes because it increases the longevity of language. We have already seen how language itself increases the fecundity and fidelity of copyable sounds; the problem was longevity. Stories told using language can be remembered in human brains but, that aside, the sounds of language are necessarily ephemeral. Writing is the first step towards creating long-lived language.

No one knows how many times writing was independently invented from scratch, but the task is formidable. To start from scratch means making a large number of decisions about how to divide up speech and how to organise the marks that are going to stand for that speech. The Sumerians of Mesopotamia invented writing about five thousand years ago; the Mexican Indians some time before 600 BC; and Egyptian and Chinese systems may also have arisen independently. Sumerian cuneiform began, like many writing systems, as an accounting system representing sheep and grain. It started with clay tokens and gradually evolved into a system of marks on clay tablets, with conventions about making the marks in order from left to right and top to bottom. Other systems, naturally enough, use different conventions. From a memetic point of view we can imagine lots of people trying out different ways of using the marks and some ways being copied more than others. This selective copying is memetic evolution at work, and the result is better and better writing systems.

Many writing systems have taken a starting point from other systems, or even just borrowed the idea of writing itself. In 1820, a Cherokee Indian called Sequoyah observed that Europeans made marks on paper and went on to devise a system for writing down the Cherokee language. Although he was illiterate and knew no English, his observations were enough for him to devise a writing system so successful that Cherokees were soon writing, reading, and printing their own books and newspapers (Diamond 1997).

I have suggested that human consciousness is not the driving force behind the creation of language (or anything else for that matter) and Sequoyah looks like the ideal case to prove me wrong. In fact, I chose him as a perfect opportunity to explain what I mean. Sequoyah was presumably as conscious as any human being. In discussions about creativity people often assume that consciousness is somehow responsible for creativity, but this view meets with serious problems as soon as you try to imagine exactly what it means. You are almost forced into adopting a dualist position, with consciousness as something separate from the brain, that magically leaps in and invents things. A more common view in science is to ignore consciousness and treat creativity as a product of the intelligence and ability of the individual concerned – ultimately taking the process back to brain mechanisms. This escapes from the dualist trap but leaves out the importance of all the ideas already available in the creator's environment. The memetic view includes all this. What I am proposing is this.

Human brains and minds are a combined product of genes and memes. As Dennett (1991, p. 207) puts it 'a human mind is itself an artifact created when memes restructure a human brain in order to make it a better habitat for memes'. In Sequoyah's case he must have had an exceptional brain, with exceptional determination and motivation, and he happened to come across a writing system that was already available at a time when his own people were in a position to take up his ideas and use them. Sequoyah's thinking was an essential part of the process, but was itself created out of the interplay between memes and genes. All this is a wonderful example of replicators creating design out of nowhere. As ever, there is really no designer other than the evolutionary process.

There are basically three strategies for writing systems. Signs can be used to stand for whole words, for syllables, or for just single sounds. The differences are important for the memes each syllable will be able to transmit. A system based on whole words is clumsy because there are so many words. Every time a new word is invented a new sign has to be created as well. At the other extreme systems using signs for single sounds can use few signs and combine them in many different ways, such as the alphabet of twenty-six letters in which this book is written. The cognitive load placed on the brains of people using the system varies in the same way. It is relatively easy to learn twenty-six letters and their sounds, although even this typically takes schoolchildren many months or even years of work. But learning Japanese *kanji* takes much longer, and unless you know two or three thousand of them you cannot read a Japanese newspaper.

For many reasons, writing systems based on sounds can convey more memes for less effort, and therefore are likely to win out in competition with other systems. Of course, the competition is not straightforward. The historical process by which writing systems are created means that there are all sorts of quirks, oddities, and arbitrary conventions which, once learned by sufficient numbers of people, attain some kind of stability. In biological evolution an important principle is that evolution always builds on what it has available at the time. There is no evolutionary God who can look at the design of the eye and say 'It would be better if we got rid of this bit and started again'. There is no starting again. The same applies to the design of writing systems. They evolve gradually from wherever they have got to at any point. So, the alphabet of twenty-six letters is far from the ideal that a memetic God would create, but it is better than many other systems, and therefore, when direct competition arises, tends to win. Many languages, like Turkish for example, have changed over from more cumbersome systems to the Roman one. Many languages use variations on the system, adding umlauts or circumflexes, diphthongs, or even new letters. We have yet to see whether the economic and cultural power of Japan is enough to ensure the survival of its complicated writing system in a world in which the transmission of memes is everything, and English written with the Roman alphabet is dominant. For memetic reasons I suspect it will not be.

A similar argument applies to numerical systems. Arithmetic is formidably hard using Roman numerals but easy with any system that relies on the *position* of a numeral, like the Arabic system that we, and most of the world, now use.

This drive towards uniformity is interesting, and is stronger than was the case for the evolution of language. In the case of writing, the invention of a new system is so difficult that borrowing one from elsewhere is more common, and novel systems are at a disadvantage. Once an adequate system has begun to evolve it has a natural advantage, in spite of any shortcomings due to historical accident and arbitrary conventions. When just a few systems exist, the one that produces slightly more, slightly better or slightly longer-lived copies begins to fill the world with its products, and the products take the idea of that copying system with them. The result is pressure towards one copying system taking over entirely from all the others.

We are all too familiar with this process. The standard QWERTY keyboard was devised to prevent the letters sticking together in the earliest manual typewriters; it is far from the optimum for modern keyboards and yet is almost universally used. Once music could be recorded and stored, vinyl disks of just two sizes and three rotation speeds captured the market, but have now mostly disappeared. Standard reel-to-reel tapes hung on for a while after the invention of the much smaller cassette tape but then cassettes persisted in a single format until the compact disk appeared, and may or may not continue to survive alongside it. Whether they do or not should be predictable from memetic principles. The number of memes that can be crammed on to a CD is dramatically greater than the number on tape, and CD technology allows rapid random access. Therefore, once cheap CD copying devices become available, CDs will surely outnumber cassettes, carrying with them the memes for that copying mechanism. The number of compact disks in the world is now so huge, not to mention the number of factories legitimately making them, and the even larger number illegally copying them, that an enormous step forward in fidelity or fecundity of copying would be needed to oust the system for a new one. The same has happened with the format of computer disks.

Bearing in mind the dangers of comparing memes and genes, we can speculate that the same process works in both cases, producing a uniform high-fidelity copying system capable of creating a potentially infinite number of products. The genes have settled down, for the most part, to an exquisitely high-fidelity digital copying system based on DNA. The memes have not yet reached such a high-quality system and will probably not settle on one for a long time yet.

Returning to writing, I have described its evolution as a step towards greater longevity for memes based on language. That step opens the way for further steps in increasing fidelity and fecundity. Spelling can vary greatly, leading to ambiguity and low fidelity. Many languages began with optional spelling that gradually gave way to 'correct' ways of spelling every word, dictionaries that specify the correct spelling and, more recently, spell checkers that enforce the rules in electronically stored text.

Fecundity is obviously limited when writing is slow and difficult, as it was for marking clay or making clay tokens to stand for words. For most of its history, writing was a skill confined to a few specially trained scribes. This made political sense because of the power it gave to rulers. They alone could command scribes to keep records of barter, financial transactions, and taxes, or to maintain holy texts for the justification of oppression and war. In any case, the early writing systems were only capable of recording limited kinds of information. It took political and economic changes, as well as changes in writing itself, before writing could be used for poetry, novels, personal letters and recording history. Widespread literacy came later with its dramatic increase in the number of memes stored and passed on as marks on paper.

The printing press was a critical step for both fecundity and fidelity. Up until the fifteenth century, all copying of texts in Europe was done by scribes, often monks who spent a large proportion of their time copying and illuminating religious works. The work was slow and they made many errors. These errors are now of great interest to historians tracking the history of texts, but they certainly did not help fidelity. The time taken meant that few copies could be made, and books were an expensive commodity for only the richest and most powerful people. This restricted the ideas in books to those for which there was financial backing – that is, ideas maintaining political, economic, and religious power. Once books were cheaply available the kinds of memes contained in them could proliferate and change. Written material is no longer confined to lists of taxes and religious tracts, but is constrained by quite different market forces. The memes took a great step forward when they got into books.

Memes in books provide a good example of a selection system at work. In this system, the replicators are the memes: the ideas, stories, theories or instructions conveyed in the printed words. These either get copied or not, and their content affects the likelihood of their being copied. The copying machinery is the publishing houses, printing presses, and factories in which the books are made. The selective environment is the minds of authors in which memes compete to get into the final text, a world full of bookshops that stock the books or not, the book reviewers and magazines that publicise the books or not, and the people who buy and read them and recommend them to their friends – or not. We humans are, obviously, critical to the whole process. However, our creative role is not that of an independent designer conjuring ideas from nowhere. Rather we are the copying machines, and parts of the selective environment, in a vast evolutionary process driven by the competition between memes.

As I write this book I think of my mind as a battleground of ideas. There are far more of them than can possibly find their way on to the final printed pages. 'I' am not an independent conscious entity creating the ideas out of nowhere. Rather, this brain has picked up millions of memes from all its education, reading, and long hours of thinking, and they are all fermenting in there as the fingers type. After this internal selective process is over and the manuscript is sent off there will be more selection, by the readers chosen by the publisher, and ultimately by the reviewers, bookshops and readers out there in the world. Whether the book sells a few hundred copies or a few hundred thousand copies will depend entirely on that selective process.

Communications

Railways, roads and ships may not seem to be directly concerned with memetic copying, but they play a role in speeding up the process of memetic competition. They carry to distant places the letters in which memes are written and the goods and people who convey ideas. They also increase the number of people who are in contact with each other which provides a larger and more varied meme pool. Just as biological evolution produces more species on large landmasses than on small islands, so memetic evolution produces more developments when more people are joined together into a memetic system. Roads, railways, and airlines connect larger and larger numbers of people together, just as common languages and writing systems do.

In a 1901 classic, *Cosmic Consciousness*, the mystic Richard Bucke predicted that with the invention of 'aerial navigation', cities would no longer be needed and rich people would live in beautiful places, evenly spread out across the globe. In fact, cities have increased dramatically in population and rural depopulation is the norm. Why is this? A memetic answer, though a slight digression from copying technology, takes a familiar form. People who live in cities meet more people and therefore pick up and pass on more memes than people who live in isolated places. Among these memes are behaviours that are only possible (or are much easier) in cities – eating out and going to pubs, going to cinemas, theatres, museums and art galleries, visiting friends at a moment's notice, or having a high-powered job at the centre of the action. The city-dweller not only picks up these memes but meets other people who also have them. Once these habits are picked up they are hard to drop.

Meanwhile, the people who live in the country meet fewer people, and do not have the opportunity to pick up the habits of exciting city life – unless they go to the city, in which case they may be lured by all the memes they find there. There is a critical imbalance operating here. When city-dwellers go to the country they meet few rural dwellers because they are widely spread out, and pick up few rural memes because few exist; but when country folk go to the city they meet lots and lots of city people and lots of new ideas. The consequence is memetic pressure for city-dwelling.

You may object that people make their choices about where to live either out of economic necessity or by freely choosing the life they know will make them happier. But is this really so? Economic necessity is often not a question of food and clothes for the family, but of buying televisions and cars and all the other trappings of a meme-rich life. The more we are exposed to memes the more we seem to acquire a hunger for them that is rarely satisfied. And happiness is very hard to judge. We may think that having a more exciting life, closer to the centre of the action, will make us happier, but we may be wrong. I suggest that we are, to a far greater extent than we would like to believe, driven to our choices by the pressure of memes.

This memetic argument suggests that there will be pressure for people to live in vast cities whenever the following conditions obtain: first, that there is enough communication between the country and the cities to set up the imbalance, and second, that people's main form of communication is still face to face, or via cheap local phone calls. If memetic transfer were truly independent of distance then the demographic pressures would change.

....

The telegraph and telephone, radio and television, are all steps towards spreading memes more effectively. They increase the fecundity of the copying process, and the distance over which it operates. People have often been unable to predict how such inventions would actually be used and which would last and which not, but from a memetic point of view prediction should be relatively easy. Anything with higher fidelity, fecundity, and longevity than its rivals should be successful. From the first electric telegraph, in 1838, to the telex machine and fax, fidelity and fecundity have gone on increasing – opening up new niches for further development along the way.

The telephone was bound to be a success. People are genetically evolved to chat and gossip (Dunbar 1996), and want to exchange news and views, creating lots of memes in the process. They can spread the memes by letters which take minutes or hours to write, and days to arrive, or they can ring each other up. People who use the phone will get more ideas spread simply because it is quicker, and those ideas include the idea of using the phone. Mobile phones have progressed very rapidly from being an executive luxury to being indispensable to every doctor, plumber, and aspiring teenager.

Letters will win out only when there is a need for longevity over fecundity. Fax machines combine the fidelity and longevity of writing with the speed (and hence fecundity) of the telephone. Photocopiers were a fantastic step for fecundity. Interestingly, people keep predicting the end of books. When radio came along predictions were made that no one would read any more. The same was proclaimed with the advent of television and then personal computers. In fact, books of a TV series can sell millions, and bookshops are selling more, not fewer, books than ever. Perhaps this is because memes can take different routes to success, just as genes do with their alternative strategies under *r*-selection and *K*-selection (p. 100). Electronic-mail messages go for high fecundity, low fidelity, and low longevity (people send out lots, do not bother to write carefully or correct the mistakes, and throw them away). Letters go for low fecundity, high fidelity, and high longevity (people write fewer letters, construct them carefully and politely, and often keep them). Books are high on all three.

All this makes a lot more sense if you look at the process as memetic competition. Any copying process that produces a successful combination of high-fidelity, long-lasting copies of memes will spread more memes and, in the process, spread itself. As this process continues more memes spread faster and faster. Note that the consequence of this is a headache for humans. Competition in business, publishing, the arts and science all depends on the transfer of memes. As memetic transfer speeds up so the competition speeds up, and people without the latest technology fail in that competition. We are driven by the latest technology to have to read all those books today, send that fax now, or be on the end of a phone line to Japan at three in the morning. We may think all this progress is designed for our own happiness, and indeed we may sometimes very much enjoy our meme-rich lives, but the real driving force behind it all is the interest of the memes.

From copy-the-product to copy-the-instruction

So far, I have talked about increasing fidelity in rather general terms. I want now to be more specific and apply two further principles to how copying systems increase their fidelity. The first is the switch from analogue to digital systems and the second the switch from copy-the-product to copy-the-instruction.

Digitising information is a good way to increase fidelity because it reduces errors in storage and transmission (p. 58). Language includes discrete words and is therefore more digital than other communications such as cries, howls, and calls. Writing extends the digitisation by committing certain sounds to certain letters, enforcing standard spellings and, above all, by allowing the vagaries of handwriting to be ignored by anyone who has learnt an alphabet. The ability of humans to read scrawly idiosyncratic handwriting is amazing, and computers are still bad at it. We are essentially able to interpret a wide variety of scribbles as being the letter 'p' or the letter 'a', thus creating a digital signal out of an analogue one. The same has been true of sound-receiving technology as it switched from grooves in disks or analogue magnetic signals stored on tape, to digital recording and storage. Indeed, it was the advent of digital sound recording that made it obvious that digital is better than analogue. Many radio stations have already changed over to entirely digital systems with a significant improvement in quality. The copying of DNA has built-in error-correction mechanisms that far exceed anything the memes have yet created.

The second step is to copy the instructions rather than the product. I previously gave the example of a recipe for soup. It may be possible for a cook to taste the soup and copy it, but the copy is likely to be better if he works from a recipe. Why? The general principle is that following recipes is not a reversible process, whether we are talking about the genetic instructions for making a body, or the recipe for a cake (Dawkins 1982). Follow the genetic instructions in the right way and under the right conditions and you get a body, but you cannot take the body and follow the instructions back to arrive at a person's genome. The same is true of the soup. Of course you can try, but errors are bound to creep into the reverse engineering required to copy the product. You have to work out how it was done, and then do it yourself. If copies of copies are made the errors are compounded, and any good tricks invested in the original product are soon lost. It is far better to have clear instructions to follow.

The invention of writing makes possible all sorts of steps in this direction. Recipes for food are only one example, others are car-maintenance manuals, instructions on how to get to the party, user manuals for hi-fi systems or gas ovens, instructions for building model aeroplanes or decorating your house in the latest fashionable styles. In these, and many other cases, you may see a product or action and guess at how it was made, but verbal or written instructions are a great help.

Copying written instructions is also far more secure. Writing is digital and highly redundant so that errors in spelling or syntax, or degradation by photocopying, are routinely ignored in passing on recipes or instructions. The same instruction can be copied to millions of people, as many computer manuals have been, and each person receives the same information. The booklet can be passed on to reader after reader without losing any detail.

I am returning to this principle because it has been so important in the computer revolution. Computer programs are instructions. They work on the basis of copy-the-instruction not copy-the-product. Take a familiar suite of programs such as the word processor I am using to write this book – Word 6.0. Word has evolved gradually through several stages and there are now millions of copies of its various versions in the world living inside millions of PCs in offices and houses. Some people buy them on disks or CDs, others copy them (legally or not) from each other. When installed the programs all do the same things. They put up letters on a screen, move text around when commanded by the user, send data to printers, and so on. No human being, from watching the word processor at work or seeing the documents it creates, could reconstruct the machine code on which it is based. The fantastic success of the memes inside Word 6.0 is due not only to its usefulness to the humans who use it, but to the digital copying machinery on which it operates and the fact that it is instructions and not products that are copied. These memes, or some of them, will outlive Word 6.0. If Word 8 or 9 is made it will doubtless reuse much of the code that formed earlier versions.

Note that the billions of products created by these word processors are not copied in the same way as the memes inside the word processor itself. But nor are they irrelevant to the copying process. If people were not happy with the program, and could not easily write all their letters, articles and books with it, then Word 6.0 would not be copied at all. It is the quality and quantity of the documents created that determines the success of the word processor they were created with. We can now see that these documents play for the memes, a similar role to that played by organisms for the genes. In this sense they are a vehicle, except that they do not carry the replicators around inside them. The documents themselves may disappear, but their existence determines which of the instructions for making them are copied and which are not. And potentially these instructions can be copied on for ever, just as genes can.

Many meme-copying steps have gone into the creation of the computers on which all this depends. They include the invention of language, its increased longevity by writing, increased communication between people by the building of roads and railways, the invention of telephones and televisions, the invention of digital computers, programming languages, digital storage devices, and finally the creation of user packages such as word processors, statistical packages, spreadsheets, and databases, which consist of memeplexes whose vehicles are the documents they make possible. We may expect this process to continue with the creation of more and more computer-based instructions whose operations are inscrutable to their users but whose products determine whether they are replicated or not.

Note that this evolutionary process has made memetic-copying mechanisms more similar to genetic ones. One of the great worries for memetics was the accusation that memes are passed on by Lamarckian 'inheritance of acquired characteristics'. We can now see that with further developments of meme-copying technology the tendency is, just as it presumably was for genes, towards a non-Lamarckian mechanism – that is, copy-the-instruction not copy-the-product. The precise way it is done will always be different for memes and genes but the basic evolutionary principles are the same. The competition between replicators forces the invention of better and better systems for copying those replicators. The best systems are digital, have effective error-correction mechanisms, and copy the instructions for making the products, rather than the products themselves.

Caught in the web

In 1989, the World Wide Web was invented. The Internet had already been expanding for many years, and what had begun as a small scheme linking a few government scientists, rapidly became a worldwide system through which anyone with a computer and modem could fetch stored information from all around the world. This was a great step for the memes. Memes can now be stored on the hard disk of a computer in, say, Melbourne, and at any time of day or night be copied almost without errors over phone lines or satellite connections to another computer in London, Florence, Chicago, or Tokyo, using the energy resources of countless human beings along the way.

These memes can be used to create other products (say school projects or business plans). They can be saved on disks at the new location, or to save space just the link be saved and the information called up again whenever needed. This last fact mirrors an interesting trick used by the human visual system. The visual world is so complex that storing even tiny fractions of the changing image would overwhelm even the vast storage system of the human brain. Instead, the brain throws away most of the information and relies on our ability simply to look again. We may have the impression that when we look out of the window we have a beautifully rich visual image, but in fact all our brains are holding is a little piece of the central image, a very rough sketch of the rest, and the ability to respond quickly to change and look again when necessary (Blackmore *et al.* 1995). In the same way, when using the Net we can mark information we might want again without actually keeping it on our own computers. The memes stay where they always were, in Sydney or Rome; and we just have a quick route for getting them again.

Use of the World Wide Web is free. This may change, but at the end of the twentieth century you pay only for the computer and phone lines that connect you to the system. Out there in cyberspace lie all the stories, and pictures, and programs, and games, that millions of people have lovingly put onto their Web sites, creating a virtual world of digital information. There are multi-user domains, or MUDS, that are imaginary places that people have constructed for others to come and play in. For some people these virtual worlds are more real than ordinary life (Turkle 1995). There are controls on who can enter a MUD but they are not financial controls. This is odd if you think of the Internet as something that humans created for their own benefit, because you might expect them to pay for it. It makes more sense when you think of the memes as having created the

Web to aid their own replication, and competing with each other to get your attention. If memes can get copied they will, and the Internet copies a lot of memes.

Does the Net need us? Yes, at the moment it does, though not necessarily for ever. We made the hardware and software on which it depends and we need to keep maintaining it, or the copying system will collapse. More importantly, our biologically evolved nature still drives, to a very large extent, which memes are successful. They are, naturally enough, those to do with sex, food, and fighting. The most common topic for searches on the World Wide Web is sex. MUDs allow people to take on invented identities and engage in meeting, chatting up, and having virtual sex with people whose location and even biological sex they may not know. The vast majority of computer games are based on killing and warfare. Any memes which can get into or tag along with such memeplexes are more likely to succeed. In this sense, the Internet still needs us, and is driven by human genes as well as memes.

However, many changes lie ahead. Already there are free-floating programs which move around in cyberspace, called bots (short for robotic programs). The way forward in artificial intelligence seems to be to build small and stupid units that together do clever things. We can imagine the Net becoming full of such autonomous stupid creatures that run about doing useful jobs. For example, as the Net increases in size and complexity, which memetic principles dictate that it must, there will be increasing problems of traffic flow and control. One idea is to create little programs based on insects laying chemical trails, that move around providing information about traffic flow on different routes. Others might perform error-correction tasks or censorship duties. At the moment, the only viruses or parasites are ones deliberately created by malicious (or just mischievous) human beings, but could bots mutate into viruses and start clogging up the system? Certainly, copying errors happen in any system and occasionally they lead to a product that proliferates. General evolutionary principles suggest that this may occur if the fantastic copying and storage system of the Net is maintained for long enough.

Other programs simulate people; they can carry on conversations and do things like psychic readings, or take part in games. There are 'chatterbots' with whom you can converse when you get lonely. In multi-user games people have been fooled by bots claiming to be real people. In a large system over a long time such bots could presumably mutate into more and more efficient 'people'.

Many people seem to assume that because we built the machinery on which the net runs, we are in control of it. This is clearly not so. British Telecom can no longer understand its own telephone network, and the whole worldwide system looks set to become bigger and more complex still. Indeed, if the memetic analysis I have given here is correct, then so long as human beings maintain the infrastructure, the system will proliferate out of anyone or anything's control – like a vast natural ecosystem.

The same applies to robots. At the moment, they mostly carry out simple tasks under human control, but memetics raises the following interesting possibility. For robots to become like humans – in other words, to have human-like artificial intelligence and artificial consciousness – they would need to have memes. Rather than being programmed to do specific tasks or even to learn from their environment as some already can, they would have to be given the ability to imitate. If they could imitate the actions of people or other robots, then robot memes would begin to spread from one to another, and a new kind of memetic evolution take off, perhaps inventing new kinds of language and communication. The robot memes would drive the robots to new activities, giving rise to motivations that we could only guess at. We humans might not be capable of imitating everything the new robots did and so we might be quite excluded from their kind of cultural evolution. We would certainly not be in control of it.

All this raises interesting, and perhaps frightening, questions about the nature of human control and human identity. In any case, memetics raises those questions from its very foundations. I have carefully avoided them so far but the time has come to ask the difficult ones. Who am I, and what am I here for?

CHAPTER 17

The ultimate memeplex

'We, alone on earth, can rebel against the tyranny of the selfish replicators'. So ends Dawkins's book *The Selfish Gene* in which the whole idea of memes began. But who is this 'we'? That is the question I want to ask now. The 'ultimate memeplex' of my title is no science fiction futuristic invention, but our own familiar self.

Think for a moment about yourself. I mean the 'real you', the inner self, that bit of yourself that really feels those heartfelt emotions, the bit of you that once (or many times) fell in love, the you that is conscious and that cares, thinks, works hard, believes, dreams and imagines; I mean who you really are. Unless you have thought about this a good deal you probably jump to many conclusions about your self – that it has some kind of continuity and persists through your life, that it is the centre of your consciousness, has memories, holds beliefs and makes the important decisions of your life.

Now I want to ask some simple questions about this 'real you'. They are: What am I? Where am I? What do I do?

What am I?

You may be one of the large majority who believes in the existence of a soul or spirit. Ethnographic studies show that most cultures include notions of a soul or spirit, nearly half believing that the soul can separate from the body (Sheils 1978). Surveys show that in the United States 88 per cent believe in a human soul, and in Europe 61 per cent, figures in line with high levels of belief in God, life after death, and supernatural phenomena (Gallup and Newport 1991; Humphrey 1995). Presumably, people assume that the soul is their inner self or 'real me' and will survive when their body dies.

There is a long history of philosophers and scientists trying to make sense of such a view. In the seventeenth century, the French philosopher René Descartes took a wonderfully sceptical view of the world, doubting every belief and opinion he had. He decided to treat everything as though it were absolutely false 'until I have encountered something which is certain, or at least, if I can do nothing else, until I have learned with certainty that there is nothing certain in the world.' (Descartes 1641, p. 102). Amidst all his doubts, he concluded that he could not doubt that he was thinking. Thus he came to his famous 'Cogito ergo sum' – I think therefore I am – and to what is now known, after him, as 'Cartesian dualism': the idea that that thinking stuff is different from physical, or extended, stuff. Our bodies may be a machine of sorts but 'we' are something else.

Dualism is tempting but false. For a start no such separate stuff can be found. If it could be found it would become part of the physical world and so not be a separate stuff at all. On the other hand if it cannot, in principle, be found by any physical measures then it is impossible to see how it could do its job of controlling the brain. How would immaterial mind and material body interact? Like Descartes' 'thinking stuff', souls, spirits and other self-like entities seem powerless to do what is demanded of them.

Nevertheless, a few scientists have developed dualist theories. The philosopher Sir Karl Popper and neuroscientist Sir John Eccles (1977), suggest that the self controls its brain by intervening at the synapses (or chemical junctions) between neurons. Yet as our understanding grows of how neurons and synapses work there is less and less need for a ghost to control the machine. Mathematician Roger Penrose (1994) and anaesthetist Stuart Hameroff (1994) suggest that consciousness operates at a quantum level in the tiny microtubules inside the membranes of neurons. Yet their proposal just replaces one mystery with another. As the philosopher Patricia Churchland (1998, p. 121) observes 'Pixie dust in the synapses is about as explanatorily powerful as quantum coherence in the microtubules'. These attempts to find a self that lurks in the gaps in our understanding just do not help, and few scientists or philosophers are convinced by them.

The opposite extreme is to identify the self with the whole brain, or whole body. This might seem appealing. After all, when you talk about Simon you mean him – the whole body, the entire person. So why not say the same of yourself? Because this does not get at the problem we are struggling with – that it feels as though there is someone inside who is consciously making the decisions. You can point to your body and say 'that is me' but you do not really mean it. Let us try a thought experiment. Imagine for a moment that you are given a choice (and you cannot say neither). Either you will have your body completely swapped for another body and keep your inner conscious self, or you will have your inner self swapped for another unspecified self and keep the body. Which will it be?

Of course, this is both practically and conceptually daft. Unless we can identify this inner self the experiment could not be done, and even then it implies a further self to do the choosing. However, the point is this. I bet you did make a choice and I bet you chose to keep your inner self. However daft the notion is, we seem to have it, and have it bad. We think of ourselves as something separate from our brains and bodies. This is what needs explaining, and so far we are not getting on very well.

This problem applies to any scientific theory that leaves the sense of self out of the picture. The most thorough-going reductionist view of this kind is what Nobel laureate Francis Crick calls 'The Astonishing Hypothesis':

> The Astonishing Hypothesis is that 'You', your joys and your sorrows, your memories and your ambitions, your sense of personal identity and free will, are in fact no more than the behavior of a vast assembly of nerve cells and their associated molecules. As Lewis Carroll's Alice might have phrased it: 'You're nothing but a pack of neurons' (Crick 1994, p. 3).

There are at least two problems with this. First, you do not feel like a pack of neurons. So what the theory needs, and does not provide, is an explanation of how a pack of neurons comes to believe that it is actually an independent conscious self. Second, the theory does not say *which* neurons. It cannot be all neurons because 'I' am not consciously aware of most of what goes on in my brain; 'I' do not identify with the neurons that control glucose levels in my blood or the fine movements that keep me sitting up straight. On the other hand if you try to identify 'self' neurons you are doomed to trouble. All neurons look much the same under the microscope and all of them are doing something all the time regardless of what 'I' am doing. Crick is working on the theory that neurons bound together by simultaneous firing at 40 cycles per second form the basis for visual awareness, but this is not the same as a theory of a conscious self.

Note that this theory is more reductionist than many others. Crick not only assumes that you are utterly dependent upon the actions of nerve cells – most neuroscientists assume that – but that you are *nothing but* the pack of neurons. Other scientists assume that new phenomena may emerge from simpler ones, and cannot be understood by understanding the underlying neurons and their connections. For example, we cannot understand human intentions, motivations, or emotions just by observing the behaviour and connections of neurons, any more than we can understand the activity of a desktop computer by looking at its chips and circuits. On this more common view the intentions depend completely on the neurons (just as the computation depends completely on the chips in the computer) but to understand them we must work at an appropriate level of explanation. But what is the appropriate level of explanation for the self? The behaviour of neurons seems to miss it.

Another approach is to identify the self with memory or personality. Victorian spiritualists believed that 'human personality' was the essence of the self and could survive physical death (Myers 1903). However, personality is nowadays understood not as a separate entity but as a fairly consistent way of behaving that makes one person identifiably different from another. This way of behaving reflects the kind of brain we were born with and our lifetime's experiences. It cannot be separated from our brain and body any more than our memories can. The more we learn about personality and memory the more obvious it is that they are functions of a living brain and inseparable from it. In an important sense you are your memories and personality – at least, you would not be the same person without them – but they are not things, or properties of a separate self. They are complex functions of neural organisation.

A final way of looking at the self is as a social construction. If I asked you who you are, you might answer with your name, your job, your relationship to other people (I'm Sally's mum or Daniel's daughter), or your reason for being where you are (I'm the cleaner, Adam invited me). All of these self-descriptions come out of your mastery of language, your interactions with other people, and the world of discourse in which you live. They are all useful in certain circumstances, but they do not describe the sort of 'inner self' we were looking for. They describe no persistent conscious entity. They are just labels for an ever-changing social creature. They depend on where you are and who you are with. We can find out a lot about how such constructions are created – indeed social psychologists do just that – but we do not find a conscious self this way. The inner 'me' seems to be mighty elusive.

Where am I?

You probably feel as though 'you' are located somewhere behind your eyes, looking out. This seems to be the most commonly imagined perspective, though others include the top of the head, the heart, or even in the neck, and there are apparently cultural differences in this imagined position. The location may change with what you are doing, and you may even be able to move it around at will. Blind people report feeling themselves in their fingertips when reading Braille, or in their long white cane when walking. Drivers sometimes inhabit the edges of their cars and wince if something passes too close. So is there anything actually at this imagined spot? Presumably not in the case of the stick or the car, but it still feels as though there is a self in there somewhere. Where, then, should we look for the self?

The most obvious place to look is in the brain. Drugs that affect the brain affect our sense of self, and damage to various areas of the brain can destroy or change it. Stimulating the brain with electrodes can produce changes in body image, feelings of shrinking or expanding, or sensations of floating and flying. Yet we do not feel as though we are inside a warm, wet, and pulsating organ. In a lurid thought experiment Dennett (1978) imagines his brain being removed to a vat in a life-support lab while his body roams around as usual, connected to his body as intricately as it ever was before, but by radio links instead of nerves. Now where would Dennett feel he was? As long as he could see and hear, he would feel as though he was wherever his eyes and ears were. He would not fancy himself to be inside the vat. Of course, we cannot do the experiment to check his intuitions, but it suggests the disturbing conclusion that Dennett would still imagine he was living in there, somewhere behind his eyes, even if the skull were empty and his brain were controlling things from the vat.

If we look inside the brain we do not see a self. To the naked eye a human brain looks like a lump of rather solid porridge with a convoluted shiny surface and various areas of paler or darker grey; it is hard to believe that all our thinking goes on in there. Only with high magnification and the techniques of modern neuroscience can we find out that it contains about a hundred billion neurons or nerve cells. The neurons are connected up in fantastically complex ways and, by virtue of these connections, store and process the information that controls our behaviour. However, there is no centre of action where a self might reside. There is no one place into which all the inputs go, and from which all the instructions get sent out. This is an important point, and deeply disturbing. We feel as though we are a central observer and controller of what goes on, but there is no place for this central controller to live.

Let us consider what happens when you perform a simple task. For example, find a letter ‘p’ on this page and then point to it. What has gone on? It may feel as though you have decided to find a ‘p’ (or not if you could not be bothered), searched the next few lines, found one, and then commanded your finger to move into position and touch it. The role of the self seems obvious, 'you' decided to act (or not), 'you' moved your finger and so on.

From an information-processing point of view the role of the 'you' is not at all obvious. Light enters the eye and is focused on a layer of light-sensitive cells. The output from these goes into four layers of cells in the retina which extract edges and brightness discontinuities, enhance differences across boundaries, change the coding of colour information from a three-receptor system to one based on pairs of opposites, and throw away a great deal of unnecessary detail. The part-digested information is then compressed and passed along the optic nerve into the thalamus inside the brain. Here, different types of information about the image are separately processed and the results passed on to other parts of the visual cortex at the back of the brain. As the information passes through it is at some times and places coded like a map, with neighbouring positions corresponding to neighbouring locations in the world, but, at other times and places, as more abstract information about shape, movement or texture. Throughout the system there are numerous things going on at once.

From the visual cortex, outputs go off to other parts of the brain, for example, those dealing with language, reading, speech, object recognition and memory. Since you know how to read, a search identifies a letter 'p'. Some of the information goes to the motor cortex which co-ordinates action. From here a movement such as pointing with your finger, will be pre-processed and then coordinated with visual feedback as it happens, so that the finger ends up in line with the 'p'.

The details of this do not matter. The important point is that the description that neuroscientists are building up of the way the brain works leaves no room for a central self. There is no single line in to a central place, nor a single line out; the whole system is massively parallel. In this description there is no need for a 'you' who decided to find the 'p' (or not) and who started the finger moving. The whole action inexorably created itself, given this book with its instruction, and your brain and body.

You might think that there is still room for a central self as some kind of informational or abstract centre rather than an actual place. There are several theories of this kind, such as Baars's (1997) global workspace theory. The workspace is like a theatre with a bright spotlight on the stage; the events in the bright spot are the only ones 'in consciousness'. But this is only a metaphor and can be a misleading one. If there is any sense to the idea of a spotlight, it is that at any time some information is being attended to – or actively processed – while other information is not.

However, this focus of activity changes continuously with the complex demands of the task we are performing. If there is a spotlight, it is one that switches on and off all over the place and can shine in several places at once; if there is a global workspace it is not located in any particular place. It cannot tell us where 'I' am.

The theatre metaphor may do more harm than good to our thinking about self and consciousness. Dennett (1991) argues that although most theorists now reject Cartesian dualism, they still secretly believe in what he calls the 'Cartesian Theatre'. They still imagine that somewhere inside our heads is a place where 'it all comes together'; where consciousness happens and we see our mental images projected on a mental screen; where we make our decisions and initiate actions; where we agonise about life, love, and meaning. The Cartesian Theatre does not exist. When sensory information comes into the brain it does not go to an inner screen where a little self is watching it. If it did, the little self would have to have little eyes and another inner screen, and so on. According to Dennett, the brain produces 'multiple drafts' of what is happening as the information flows through its parallel networks. One of these drafts comes to be the verbal story we tell ourselves, which includes the idea that there is an author of the story, or a user of the brain's virtual machine. Dennett calls this the 'benign user illusion'. So maybe this is all we are; a centre of narrative gravity; a story about a persisting self who does things, feels things and makes decisions; a benign user illusion. And illusions do not have locations.

What do I do?

Hold out your arm in front of you and then, whenever you feel like it, spontaneously and of your own free will, flex your wrist. You might like to do this a few times, making sure you do it as consciously and spontaneously as you can. You will probably experience some kind of inner dialogue or decision process in which you hold back from doing anything, and then decide to act. Now ask yourself, what began the process that led to the action? Was it you?

This task formed the basis of some fascinating experiments carried out by the neurosurgeon Benjamin Libet (1985). His subjects had electrodes on their wrists to pick up the action, and electrodes on their scalps to measure brain waves, and they watched a revolving spot on a clock face. As well as spontaneously flexing their wrists they were asked to note exactly where the spot was when they decided to act. Libet was therefore timing three things: the start of the action, the moment of the decision to act, and the start of a particular brain wave pattern called a readiness potential. This pattern is seen just before any complex action, and is associated with the brain planning the series of movements to be carried out. The question was, which would come first, the decision to act or the readiness potential?

If you are a dualist you may think that the decision to act must come first. In fact what Libet found was that the readiness potential began about 550 milliseconds (just over half a second) before the action, and the decision to act about 200 milliseconds (about one-fifth of a second) before the action. In other words, the decision to act was not the starting point – a finding that can seem a little threatening to our sense of self. There was much controversy over his results and many criticisms of the experiments, but given all I have said above, his results were only to be expected. There is no separate self jumping into the synapses and starting things off. My brain does not need me.

So what does my self do? Surely it must at least be the centre of my awareness; the thing that receives impressions as I go about my life? Not necessarily. This false view is just part of Dennett's illusory Cartesian Theatre. You can think about this either logically, or from the point of view of your own experience. We have already considered the logic; so now let us try to introspect carefully. Sit down comfortably and look at something uninteresting. Now concentrate on feeling the sensations from your body and on hearing what is going on around you. Stay like that long enough to get used to it and then ask yourself some questions. Where is that sound? Is it inside my head or over there? If it's over there, then what is hearing it? Can I be conscious of the thing that is hearing it? If so, am I separate from that thing as well?

You can make up your own questions. The general idea is an old one, and has been used in many meditation traditions over the millennia. Staring determinedly into your own experience does not reveal a solid world observed by a persisting self but simply a stream of ever-changing experience, with no obvious separation between observed and observer. The eighteenth-century Scottish philosopher David Hume explained that whenever he entered most intimately into himself he always stumbled upon some particular perception – of heat or cold or pain or pleasure. He could never catch *himself* without a perception, nor observe anything but the perception. He concluded that the self was no more than a 'bundle of sensations' (Hume 1739–40). The very natural idea that 'I' hear the sounds, feel the sensations, or see the world may be false.

Another series of experiments by Libet (1981) adds an interesting twist to the argument. Conscious sensory impressions can be induced by stimulating the brain, but only when it is continuously stimulated for about half a second. It is as though consciousness takes some time to build up. This would lead to the odd idea that our conscious appreciation of the world lags behind the events, but because of a process Libet calls 'subjective antedating' we never realise it is lagging behind. The story we tell ourselves puts events in order. Further experiments showed that with short stimuli (too short to induce conscious sensation) people could nevertheless guess correctly whether they were being stimulated or not (Libet *et al.* 1991). In other words they could make correct responses without awareness. Again the implication is that consciousness does not direct the action. Conscious awareness comes all right, but not in time. The hand is removed from the flame before we consciously feel the pain. We have whacked the tennis ball back before we can be conscious of it coming towards us. We have avoided the puddle before we were conscious of its existence. Consciousness follows on later. Yet we still feel that 'I' consciously did these things.

Something else we think we do is to believe things. Because of our beliefs we argue vehemently over dinner that President Clinton really could not have done it, that the Israelis ought (or ought not) to have built those homes, that private education ought to be abolished, or that all drugs should be legalised. We are so convinced of our belief in God that we will argue for hours (or perhaps even go to war or lay down our life for Him). We are so convinced by the alternative therapy that helped *me* that we force its claims on all our friends. But what does it mean to say that I believe? It sounds as though there must be a self in there who has things called beliefs, but from another perspective there is only a person arguing, a brain processing the information, memes being copied or not. We cannot actually find either the beliefs or the self who believes.

The same can be said of memory. We speak as though the self pulls up memories at will from its personal store. We conveniently ignore the fact that memories are ever-changing mental constructions, that often we fail to remember accurately, that some memories come unbidden and that we often use complex memories with no conscious awareness at all. It is more accurate to say that we are just human beings doing complex things that need memory and who then construct a story about a self who does the remembering.

In this, and many other ways, we seem to have an enormous desire to describe ourselves (falsely) as a self in control of 'our' lives. The British psychologist Guy Claxton suggests that what we take for self control is just a more or less successful attempt at prediction. Much of the time our predictions about what we will do next are reasonably accurate and we can get away with saying 'I did this' or 'I intended to do that'. When they go wrong we just bluff. And we use some truly outrageous tricks to maintain the illusion.

> I meant to keep my cool but I just couldn't. I'm supposed not to eat pork but I forgot. I'd decided on an early night but somehow here we are in Piccadilly Circus at four a.m. with silly hats and a bottle of wine . . . If all else fails – and this is a truly audacious sleight of hand – we can reinterpret our failure of control as an actual success! 'I changed my mind,' we say (Claxton 1986, p. 59).

Claxton concludes that consciousness is 'a mechanism for constructing dubious stories whose purpose is to defend a superfluous and inaccurate sense of self' (1994, p. 150). Our error is to think of the self as separate, persistent, and autonomous. Like Dennett, Claxton thinks that the self is really only a *story* about a self. The inner self who does things is an illusion.

The function of a self

Where have we got to in this brief exploration of the nature of self and consciousness? I can summarise by comparing two major kinds of theory about the self. On the one hand are what we might call 'real self' theories. They treat the self as a persistent entity that lasts a lifetime, is separate from the brain and from the world around, has memories and beliefs, initiates actions, experiences the world, and makes decisions. On the other hand are what we might call 'illusory self' theories. They liken the self to a bundle of thoughts, sensations, and experiences tied together by a common history (Hume 1739–40; Parfit 1987), or a series of pearls on a string (Strawson 1997). On these theories, the illusion of continuity and separateness is provided by a story the brain tells, or a fantasy it weaves.

Everyday experience, ordinary speech and 'common sense' are all in favour of the 'real self', while logic and evidence (and more disciplined experience) are on the side of the 'illusory self'. I prefer logic and evidence and therefore prefer to accept some version of the idea that the continuous, persistent and autonomous self is an illusion. I am just a story about a me who is writing a book. When the word 'I' appears in this book, it is a convention that both you and I understand, but it does not refer to a persistent, conscious, inner being behind the words.

Now, having accepted that, a new question arises. Why do we humans tell this story? If no persistent conscious self exists, why do people believe it does? How is it that people routinely live their lives as a lie?

The most obvious kind of explanation to try is that having a sense of self benefits the replication of our genes. Crook (1980) argues that self-consciousness arose from using Machiavellian Intelligence and reciprocal altruism, with its need for balancing the trust and distrust of others. In a rather dualistic version of a similar theory Humphrey (1986) suggests that consciousness is like an inner eye observing the brain. As primates developed ever more complex social structures, their survival began to depend on more sophisticated ways of predicting and outwitting others' behaviour. In this, he argues, *Homo psychologicus* would win out. Imagine a male who wanted to steal a mate from his rival or get more than his fair share of a kill. Predicting what the rival would do next would help, and one way to predict what others will do is to observe your own inner processes. These and other theories suggest that a complex social life makes it necessary to have a sense of self, to tot up scores in reciprocation, and to develop what psychologists now call a 'theory of mind' – that is, the understanding that other people have intentions, beliefs, and points of view.

However, this does not explain why our theory of mind is so wrong. Surely one could understand one's own behaviour without creating the idea of a separate and persistent self when it does not exist. Crook and Humphrey jump from the idea that early hominids might have benefited genetically by having an accurate model of their own behaviour to the idea that they would therefore acquire the idea of a separate self. Our self, the self we are trying to understand, is not just a model of how our own body – and by inference other bodies – is likely to behave, but a false story about an inner self who believes things, does things, wants things and persists throughout life.

Self-deception can have benefits. According to Trivers' (1985) theory of adaptive self-deception, hiding intentions from oneself may be the best way to hide them from others, and so deceive them. However, this theory does not help in the case of inventing a central self. Dennett (1991) describes us as adopting 'the intentional stance'; that is, we behave 'as if' other people (and sometimes animals, plants, toys and computers) have intentions, desires, beliefs, and so on. He argues that this metaphor of agency is a practical necessity of life; it gives us new and useful tools for thinking with. The problem is, it seems to me, that we apply this intentional stance too thoroughly to ourselves – we fall too deeply into the 'benign user illusion'. We do not say to ourselves 'it's *as if* I have intentions, beliefs and desires' but 'I really do'. I am left wondering how we get from the evolutionary advantage of having a theory of mind, or the practical advantage of adopting the intentional stance, to living our lives as a lie, protecting our ideas, convincing others of our beliefs, and caring so much about an inner self who does not exist.

Perhaps we create and protect a complex self because it makes us happy. But does it? Acquiring money, admiration, and fame gives some kind of happiness, but it is typically brief. Happiness has been found to depend more on having a life that matches your skills to what you are doing than to having a rich lifestyle. The Chicago psychologist Mihaly Csikszentmihalyi (1990) studied the fulfilling experience of 'flow' that artists describe when they lose themselves in their work. 'Flow' comes to children playing games, people deep in conversation, people skiing or mountain climbing, playing golf or making love. These all entail the same sense of happiness through loss of self-consciousness.

What makes *you* happy? Or consider the reverse: What makes *you* unhappy? Probably it is things like disappointment, fear of the future, worry about loved ones, not having enough money, people not liking you, living too stressful a life, and so on. Many of these things are only relevant to a creature that has self-awareness and the idea of a self as the owner of experience. Other animals can show disappointment, as when food does not arrive when they expect it, but they cannot have the deep disappointment of not getting a job, the fear of being thought stupid, or the misery of thinking someone they care about does not like them. We construct many of our miseries out of the idea of a persistent self that we desperately want to be loved, successful, admired, right about everything, and happy.

According to many traditions this false sense of self is precisely the root of all suffering. This idea is probably clearest in Buddhism with the doctrine of *anatta* or no self. This does not mean that there is no body, nor that there is literally no self at all, but that the self is a temporary construction, an idea or story about a self. In a famous speech, the Buddha told the monks 'actions do exist, and also their consequences, but the person that acts does not' (Parfit 1987). He taught that because we have the wrong idea about our self, we think that we will be happy if we gain more material things, or status or power. In fact it is wanting some things and being averse to others that makes us unhappy. If only we could realise our true nature then we would be free of suffering because we would know there is no 'me' to suffer.

Now we can see the difference between Dennett's view and the Buddhist one. Both understand the self to be some kind of story or illusion, but for Dennett it is a 'benign user illusion' and even a life-enhancing illusion, while for the Buddhist it is the root of human suffering. Either way it is an untruth. There is no doubt that having a clear sense of identity, a positive self-image and good self-esteem are associated with psychological health, but this is all about comparing a positive sense of self with a negative one. When we ask what good is done by having a sense of self at all, the answer is not obvious.

The selfplex

Memetics provides a new way of looking at the self. The self is a vast memeplex – perhaps the most insidious and pervasive memeplex of all. I shall call it the 'selfplex'. The selfplex permeates all our experience and all our thinking so that we are unable to see it clearly for what it is – a bunch of memes. It comes about because our brains provide the ideal machinery on which to construct it, and our society provides the selective environment in which it thrives.

As we have seen, memeplexes are groups of memes that come together for mutual advantage. The memes inside a memeplex survive better as part of the group than they would on their own. Once they have got together they form a self-organising, self-protecting structure that welcomes and protects other memes that are compatible with the group, and repels memes that are not. In a purely informational sense a memeplex can be imagined as having a kind of boundary or filter that divides it from the outside world. We have already considered how religions, cults, and ideologies works as memeplexes; we can now consider how the selfplex works.

Imagine two memes. The first concerns some esoteric points of astrology: that the fire element in Leo indicates vitality and power, while Mars in the first house indicates an aggressive personality, and transits of Mars should be ignored unless the aspect is a conjunction. The other meme is a personal belief – 'I believe that the fire element in Leo . . .' Which meme will fare better in the competition to get into as many brains, books and television programmes as possible? The second will. A piece of information on its own may be passed on if it is relevant to a particular conversation, or useful for some purpose, but it is just as likely to be forgotten. On the other hand, people will press their beliefs and opinions on other people for no very good reason at all and, on occasion, fight very hard to convince others about them.

Take another example: the idea of sex differences in ability. As an abstract idea (or isolated meme) this is unlikely to be a winner. But get it into the form 'I believe that boys and girls are equally good at everything' and it suddenly has the enormous weight of 'self' behind it. 'I' will fight for this idea as though I were being threatened. I might argue with friends, write opinion pieces, or even go on marches. The meme is safe inside the haven of 'self', even in the face of evidence against it. 'My' ideas are protected by the behaviour they induce.

This suggests that memes can gain an advantage by becoming associated with a person's self concept. It does not matter how they do this – whether by raising strong emotions, by being especially compatible with memes already in place, or by providing a sense of power or attractiveness – they will fare better than other memes. These successful memes will more often be passed on, we will all come across them and so we, too, will get infected with self-enhancing memes. In this way our selfplexes are all strengthened.

Note that we do not have to agree with or like the memes we pass on, but only to engage with them in some way. Whether it's eating pasta, watching *The Simpsons*, or listening to jazz, the memes are passed on not just in eating the food or playing the music but in statements such as 'I like . . .' 'I hate . . .' 'I can't stand . . .' Pyper concludes that 'Dawkins himself has become a "survival machine" for the bible, a "meme nest" for its dispersed memes which may induce readers who would otherwise leave their bibles unread to go back to the text' (Pyper 1998, pp. 86–7). Presumably, Dawkins did not intend to encourage religious memes in this way but his powerful response to religion has had that effect. Memes that provoke no response fare poorly, while those that provoke emotional arguments can induce their carrier to pass them on. By acquiring the status of a personal belief a meme gets a big advantage. Ideas that can get inside a self – that is, become 'my' ideas, or 'my' opinions, are winners.

Then there are possessions. Some other animals, without memes, might be said to have possessions: a robin owns the territory he guards, a powerful male owns his harem of females, and a lioness owns her kill. Human possessions can serve similar functions, such as enhancing personal status and providing a genetic advantage. But we should not overlook a big difference, that our possessions seem to belong to the mythical 'I', not just to the body it supposedly inhabits. Think of something you own and care about, something you would be sorry to lose, and ask yourself who or what actually owns it. Is it sufficient to say that your body does? Or are you tempted to think that it is the inner conscious you who owns it? I am. I realise, with some dismay, that I am partly defined by my house and garden, my bicycle, my thousands of books, my computer, and my favourite pictures. I am not just a living creature, but all these things as well; and they are things that would not exist without memes and would not matter without 'me'.

An interesting consequence of all this is that beliefs, opinions, possessions and personal preferences all bolster the idea that there is a believer or owner behind them. The more you take sides, get involved, argue your case, protect your possessions, and have strong opinions, the more you strengthen the false idea that there is not only a person (body and brain) talking, but an inner self with esoteric things called beliefs. The self is a great protector of memes, and the more complex the memetic society in which a person lives, the more memes there are fighting to get inside the protection of the self.

As the number of memes we all come across increases, so there are more and more chances for memes to provoke strong reactions and get passed on again. The stakes are thereby raised, and memes must become ever more provocative to compete. The consequence is that stress levels increase as we are bombarded by memes that have successfully provoked other people. We acquire more and more knowledge, opinions, and beliefs of our own, and in the process become more and more convinced that there is a real self at the centre of it all.

There is no 'I' who 'holds' the opinions. There is a body that says 'I believe in being nice to people', and a body that is (or is not) nice to people. There is a brain that can store knowledge of astrology and the tendency to talk about it, but there is not *in addition* a self who 'has' the belief. There is a biological creature who eats yoghurt every day but there is not *in addition* a self inside who loves yoghurt. As the memosphere becomes more and more complicated, selves follow suit. To function in our society we are all expected to hold opinions on science, politics, the weather, and relationships; to hold down a job, bring up a family, read the paper, and enjoy our leisure time. With constant memetic bombardment our lives and our selves become more and more stressful and complicated. But this is a 'Red Queen' process. No one benefits because everyone has to keep running just to stay in the same place. I wonder just how much memetic pressure selfplexes can take before they blow apart, become unstable, or divide into fragments. The unhappiness, desperation, and psychological ill-health of many modern people may reveal just this. Today's psychotherapy is a kind of memetic engineering, but it is not based on sound memetic principles. That is something for the future.

In conclusion, the selfplex is successful not because it is true or good or beautiful; nor because it helps our genes; nor because it makes us happy. It is successful because the memes that get inside it persuade us (those poor overstretched physical systems) to work for their propagation. What a clever trick. That is, I suggest, why we all live our lives as a lie, and sometimes a desperately unhappy and confused lie. The memes have made us do it – because a 'self' aids their replication.

CHAPTER 18

Out of the meme race

Now we have a radically new idea of who we are. Each of us is a massive memeplex running on the physical machinery of a human body and brain – a meme machine. Crick was wrong. We are not 'nothing but a pack of neurons'; we are a pack of memes too. And without understanding the pack of memes we can never understand ourselves.

The sociobiologists have missed a crucial point. Their achievement is to explain much of human behaviour in terms of the past selection of genes; to apply Darwin's great theory to psychology. But in concentrating on genes alone they miss out on the importance and power of the social world. To stick to their Darwinian framework they have to treat all of culture as part of the environment of genetic selection, and so they fail to see that it has its own evolutionary processes and its own power to effect change. Without the concept of the second replicator sociobiology must always remain impoverished.

By contrast, sociologists have long realised the power of social forces. As Karl Marx (1904, p. 11) argued 'It is not the consciousness of men that determines their existence, but, on the contrary, their social existence determines their consciousness.' Social scientists study the way that people's lives and selves are constructed by their roles, and by the texts in which they are embedded. But they have no evolutionary theory within which to understand the processes going on. For them the biological world and the social world are explained in entirely different ways and must remain divorced. Only when we see a human being as a product of both natural and memetic selection can we bring all aspects of our lives together within one theoretical framework.

What I am saying about human nature is so easy to misunderstand that I want to spell it out very carefully.

We humans are simultaneously two kinds of thing: meme machines and selves. First, we are objectively individual creatures of flesh and blood. Our bodies and brains have been designed by natural selection acting on both genes and memes over a long period of evolution. Although each of us is unique, the genes themselves have all come from previous creatures and will, if we reproduce, go on into future creatures. In addition, because of our skill with language and our memetic environment, we are all repositories of vast numbers of memes, some of them simply pieces of stored information, others organised into self-protecting memeplexes. The memes themselves have come from other people and will, if we speak and write and communicate, go on into yet more people. We are the temporary conglomerations of all these replicators and their products in a given environment.

Then there is the self we think we are. Among all these memeplexes is an especially potent one based around the idea of an inner self. Each selfplex has been put together by the processes of memetic evolution acting in the relatively short period of one human lifetime. 'I' am the product of all the memes that have successfully got themselves inside this selfplex – whether because my genes have provided the sort of brain that is particularly conducive to them, or because they have some selective advantage over other memes in my memetic environment, or both. Each illusory self is a construct of the memetic world in which it successfully competes. Each selfplex gives rise to ordinary human consciousness based on the false idea that there is someone inside who is in charge.

The ways we behave, the choices we make, and the things we say are all a result of this complex structure: a set of memeplexes (including the powerful selfplex) running on a biologically constructed system. The driving force behind everything that happens is replicator power. Genes fight it out to get into the next generation, and in the process biological design comes about. Memes fight it out to get passed on into another brain or book or object, and in the process cultural and mental design comes about. There is no need for any other source of design power. There is no need to call on the creative 'power of consciousness', for consciousness has no power. There is no need to invent the idea of free will. Free will, like the self who 'has' it, is an illusion. Terrifying as this thought seems, I suggest it is true.

Free will

Benjamin chose cornflakes this morning for breakfast. Why? He did so because he is a human with human tastes and the genetic make-up that inclines him towards carbohydrates in the morning, especially this morning when he was rather hungry. He lives in a rich society where cornflakes have been invented and he has enough money to buy them. He responds positively to the picture on the packet and the advertisements he sees. Memes and genes together produced this behaviour in this environment. If asked, Benjamin will say that he chose the cornflakes because he likes them, or that he made a conscious decision to eat them today. But this explanation adds nothing. It is just a story Benjamin tells after the fact.

So does Benjamin have free will or not? The critical question to ask is who do you mean by Benjamin? If by 'Benjamin' you mean a body and brain, then certainly Benjamin had a choice. Human beings make decisions all the time. Like frogs, cats, and even robots, they have plans, desires, and aversions, and they act accordingly. The more memes they acquire the cleverer are the things they can do, and the larger the range of options. They can find themselves in situations in which they have many potential choices, or few, or none. Is this sufficient for what we call free will?

I think not, because at the heart of the concept of free will lies the idea that it must be Benjamin's conscious self who made the decision. When we think of free will we imagine that 'I' have it, not that this whole conglomeration of body and brain has it. Free will is when 'I' consciously, freely, and deliberately decide to do something, and do it. In other words 'I' must be the agent for it to count as free will.

But if the memetic view I have been proposing here is right, then this is nonsense, because the self that is supposed to have free will is just a story that forms part of a vast memeplex, and a false story at that. On this view, all human actions, whether conscious or not, come from complex interactions between memes, genes and all their products, in complicated environments. The self is not the initiator of actions, it does not 'have' consciousness, and it does not 'do' the deliberating. There is no truth in the idea of an inner self inside my body that controls the body and is conscious. Since this is false, so is the idea of my conscious self having free will.

Dennett (1984) has described many versions of the idea of free will and argues that some of them are worth wanting. Unlike Dennett I neither think the 'user illusion' is benign, nor do I want any version of free will that ascribes it to a self who does not exist.

Consciousness

I have no grand theory of consciousness to offer. Indeed, the term is used in so many contradictory ways that it is hard to know what such a theory would have to accomplish. Nevertheless, I do not view the whole attempt as hopeless, as does Pinker (1998), nor as a 'hard problem' of quite a different order from any other scientific problem as does Chalmers (1996). I even think the theory of memetics may help.

First, by consciousness I mean subjectivity – what it's like being me now (p. 2). This subjectivity comes about in ways we do not understand, yet we do know that it depends critically on what the brain is doing at any time. We can look at it this way – the quality of my consciousness at any time depends on what the whole brain is doing, but particularly on the way the brain's processing resources are divided up, and the stories that are being constructed about who is doing what. In our normal state of consciousness the whole experience is dominated by the selfplex which uses words and other useful memetic constructs to weave a very fine tale. It sets everything in the context of a self who is doing things. However, when gazing in awe at the view from a mountain top, or engrossed in a creative task, the selfplex does not dominate and other states of consciousness are possible. Then there can be consciousness without self-consciousness.

Note that here my view departs from Dennett's. For him 'Human consciousness is *itself* a huge complex of memes (or more exactly, meme-effects in brains)' (Dennett 1991, p. 210). This means that a person is conscious by virtue of having all the thinking tools that memes provide, including the 'benign user illusion' and all the self memes, and without them they would, presumably, cease to have 'human consciousness'. By contrast, I suggest that the user illusion obscures and distorts consciousness. Ordinary human consciousness is indeed constrained by the selfplex, but it does not have to be. There are other ways of being conscious.

There are implications here for artificial consciousness and for animals. If ordinary human consciousness is entirely dominated by the selfplex then only systems that have a selfplex can be conscious in that way. So, since other animals do not generally imitate and cannot have memes, they cannot have the human kind of self-consciousness. This does not, however, rule out the possibility that there is something it is like to be a bat, or a rat, or even a robot.

Second, I want to emphasise that consciousness cannot *do* anything. The subjectivity, the 'what it's like to be me now' is not a force, or a causal agent, that can make things happen. When Benjamin poured out his cornflakes he may have been conscious, but the consciousness played no role in making him do it. The consciousness simply arose as what it was like to be that human being, taking those decisions, and doing those actions, and with a memeplex inside saying 'I am doing this'. Benjamin may think that if 'he' did not consciously make the decision then it would not happen. I say he would be wrong.

Critics of the analogy between genes and memes often argue that biological evolution is not consciously directed, whereas social evolution is. Even proponents of memetics sometimes make the same distinction, saying for example that 'much cultural and social variation is consciously guided in a way that genetic variation is not' (Runciman 1998, p. 177). My colleague Nick Rose (1998) accuses these theorists of 'self-centred selectionism', a mistake equivalent to the idea of directed evolution in biology. The whole point about evolutionary theory is that you do not need anyone to direct it, least of all *consciously*. When human beings act, our actions have effects on memetic selection, but this is not because we were conscious. Indeed, the most mindless and least conscious of our actions can be imitated just as easily as our most conscious ones. Cultural and social variation is guided by the replicators and their environment, not by something separate from them all called consciousness.

Creativity

Tamarisk has written a science book. This suggests that she consciously authored the book, but there is another way of looking at it. Tamarisk is a gifted writer because the genes have created a brain that handles language well, and a determined individual who likes solitary work; because she was born into a society that values books and pays for them; because her education gave her the opportunity to discover how good she was at science; and because she has spent years studying and thinking until new ideas came out of the combinations of the old. When the book was completed it formed a new complex of memes: variations on old ones and new combinations created by the complicated processes inside a clever thinking brain. When asked, Tamarisk might say that she consciously and deliberately invented every word herself (though she is quite likely to say that she has no idea how she did it). I would say that the book was a combined product of the genes and memes playing out their competition in Tamarisk's life.

This view of creativity is alien to many people. In discussions of consciousness it is common to raise the issue of creativity, as though it somehow epitomises the power of human consciousness. How could we create great music, inspiring cathedrals, moving poems, or stunning paintings unless we have consciousness? – people ask. This view of creativity betrays a commitment to a false theory of self and consciousness, or to Dennett's Cartesian Theatre (p. 225). If you believe that you live inside your head and direct operations, then creative acts can seem especially good examples of things that 'you' have done. But, as we have seen, this view of self does not hold up. There is no one inside there to do the doing – other than a bunch of memes.

I am not saying that there is no creativity. New books are written, new technologies invented, new gardens laid out, and new films produced. But the generative power behind this creativity is the competition between replicators, not a magical, out-of-nowhere power such as consciousness is often said to be. The creative achievements of human culture are the products of memetic evolution, just as the creative achievements of the biological world are the products of genetic evolution. Replicator power is the only design process we know of that can do the job, and it does it. We do not need conscious human selves messing about in there as well.

Of course selves are not irrelevant. Far from it. By virtue of their organisation and persistence, selfplexes are powerful memetic entities that affect the behaviour of the people who sustain them, and of all those who come into contact with them. But as far as creativity is concerned selves can often do more harm than good, for creative acts often come about in a state of selflessness, or loss of self-consciousness, when the self seems to be out of the way. Artists, writers and runners often say they are at their best when acting spontaneously and without self-consciousness. So selves have effects but not as the originators of conscious creativity.

Human foresight

Humans are often credited with having *real* foresight, in distinction to the rest of biology which does not. For example, Dawkins compares the 'blind watchmaker' of natural selection with the real human one. 'A true watchmaker has foresight: he designs his cogs and springs, and plans their interconnections, with a future purpose in his mind's eye. Natural selection . . . has no purpose in mind' (Dawkins 1986, p. 5). I think this distinction is wrong.

There is no denying that the human watchmaker is different from the natural one. We humans, by virtue of having memes, can think about cogs, and wheels, and keeping time, in a way that animals cannot. Memes are the mind tools with which we do it. But what memetics shows us is that the processes underlying the two kinds of design are essentially the same. They are both evolutionary processes that give rise to design through selection, and in the process they produce what looks like foresight.

As Plotkin (1993) points out, knowledge (whether in humans, animals, or plants) is a kind of adaptation. So is foresight. When a daffodil bulb starts into growth it is predicting the summer ahead, but we know this prediction was a result of past selection. When a cat predicts which way a mouse will go and pounces at the right moment we know that the ability to behave that way was naturally selected. Both these creatures have foresight of a kind, even though their genes did not. When a person predicts what she will do tomorrow or designs a new computer we somehow think this is different. The difference may seem a large one; for there is a much cleverer brain making the predictions, and the predictions may be much more complicated and precise, such as predicting the exact time of high tide or the moment when an asteroid will hit the earth. However, this kind of foresight also comes about by selection, only in this case it is selection between memes. There is no magical conscious mind that 'really' has some other kind of foresight.

The ultimate rebellion

Where does this leave us with respect to Dawkins's claim that 'We, alone on earth, can rebel against the tyranny of the selfish replicators'. Dawkins is not alone in taking the view that there is someone or something inside us who can step out of the evolutionary process and take it over.

Csikszentmihalyi (1993) explains how memes evolve independently of the people who nurture them; how the memes of weapons, alcohol and drugs are successful while doing us no good. He describes the artist not as originator but as the medium through which artworks evolve. Yet his final message is that we must take conscious control of our lives and begin directing evolution towards a more harmonious future. 'If you achieve control over your mind, your desires, and your actions, you are likely to increase order around you. If you let them be controlled by genes and memes, you are missing the opportunity to be yourself.' (Csikszentmihalyi 1993, p. 290).

In his book *Virus of the Mind*, Brodie exhorts us to 'consciously choose your own memetic programming to better serve whatever purpose you choose, upon reflection, to have for your life.' and says of the memes 'you get to choose whether programming yourself with them aids or hinders your life purpose' (Brodie 1996, pp. 53, 188).

But this is all a cop out. As Dennett says 'The "independent" mind struggling to protect itself from alien and dangerous memes is a myth' (1995, p. 365). So we must ask *who* gets to choose? If we take memetics seriously then the 'me' that could do the choosing is itself a memetic construct: a fluid and ever-changing group of memes installed in a complicated meme machine. The choices made will all be a product of my genetic and memetic history in a given environment, not of some separate self that can 'have' a life purpose and overrule the memes that make it up.

This is the power and beauty of memetics: it allows us to see how human lives, language, and creativity all come about through the same kind of replicator power as did design in the biological world. The replicators are different, but the process is the same. We once thought that biological design needed a creator, but we now know that natural selection can do all the designing on its own. Similarly, we once thought that human design required a conscious designer inside us, but we now know that memetic selection can do it on its own. We once thought that design required foresight and a plan, but we now know that natural selection can build creatures that look as though they were built to plan when in fact there was none. If we take memetics seriously there is no room for anyone or anything to jump into the evolutionary process and stop it, direct it, or do anything to it. There is just the evolutionary process of genes and memes playing itself endlessly out – and no one watching.

What then am I to do? I feel as though I have to make a choice – to decide how to live my life in the light of my scientific understanding. But how do I do that if I am nothing but a temporary conglomeration of genes, phenotype, memes, and memeplexes. If there is no choice, how am I to choose?

Some scientists prefer to keep their scientific ideas and their ordinary lives separate. Some can be biologists all week and go to church on Sunday, or be physicists all their life and believe they will go to heaven. But I cannot divorce my science from the way I live my life. If my understanding of human nature is that there is no conscious self inside then I must live that way – otherwise this is a vain and lifeless theory of human nature. But how can 'I' live as though I do not exist, and who would be choosing to do so?

One trick is to concentrate on the present moment – all the time – letting go of any thoughts that come up. This kind of 'meme-weeding' requires a great concentration but is most interesting in its effect. If you can concentrate for a few minutes at a time, you will begin to see that in any moment there is no observing self. Suppose you sit and look out of the window. Ideas will come up but these are all past- and future-oriented; so let them go, come back to the present. Just notice what is happening. The mind leaps to label objects with words, but these words take time and are not really in the present. So let them go too. With a lot of practice the world looks different; the idea of a series of events gives way to nothing but change, and the idea of a self who is viewing the scene seems to fall away.

Another way is to pay attention to everything equally. This is an odd practice because things begin to lose their 'thingness' and become just changes. Also, it throws up the question of who is paying attention (Blackmore 1995). What becomes obvious, in doing this task, is that attention is always being manipulated by things outside yourself rather than controlled by you. The longer you can sit still and attend to everything, the more obvious it becomes that attention is dragged away by sounds, movements, and most of all thoughts that seem to come from nowhere. These are the memes fighting it out to grab the information-processing resources of the brain they might use for their propagation. Things that worry you, opinions that you hold, things you want to say to someone, or wish you hadn't – these all come and grab the attention. The practice of paying equal attention to everything disarms them and makes it obvious that you never did control the attention; it controlled – and created – you.

These kinds of practices begin to wear away at the false self. In the present moment, attending equally to everything, there is no distinction between myself and the things happening. It is only when 'I' want something, respond to something, believe something, decide to do something, that 'I' suddenly appear. This can be seen directly through experience with enough practice at just being.

This insight is perfectly compatible with memetics. In most people the selfplex is constantly being reinforced. Everything that happens is referred to the self, sensations are referred to the observing self, shifts of attention are attributed to the self, decisions are described as being made by the self, and so on. All this reconfirms and sustains the selfplex, and the result is a quality of consciousness dominated by the sense of 'I' in the middle – me in charge, me responsible, me suffering. The effect of one-pointed concentration is to stop the processes that feed the selfplex. Learning to pay attention to everything equally stops self-related memes from grabbing the attention; learning to be fully in the present moment stops speculation about the past and future of the mythical 'I'. These are tricks that help a human person (body, brain and memes) to drop the false ideas of the selfplex. The quality of consciousness then changes to become open, and spacious, and free of self. The effect is like waking up from a state of confusion – or waking from the meme dream (Blackmore in press).

This kind of concentration is not easily learned. Some people are naturals and can do it relatively quickly, but for most people it takes many years of practice. One of the problems is motivation – it is hard to practise consistently just because someone else tells you this is a better way to live. This is where science can help. If our scientific understanding of human nature leads us to doubt the inner self, the soul, the divine creator, or life after death, that doubt can provide the motivation to look directly into experience; to try living without a false sense of self or false hope. Science and spirituality are often opposed but they should not be.

I have described these practices as being done for a few minutes while sitting quietly, but can all of life be lived that way? I think so, but the results are somewhat unnerving. If I genuinely believe that there is no 'I' inside, with free will and conscious deliberate choice, then how do I decide what to do? The answer is to have faith in the memetic view; to accept that the selection of genes and memes will determine the action and there is no need for an extra 'me' to get involved. To live honestly, I must just get out of the way and allow decisions to make themselves.

I say the result is unnerving because at first it is odd to observe that actions happen whether or not 'I' will them. I used to have two possible routes home, the main road and the prettier but slower lanes. As I drove up to the junction I was often torn by indecisiveness. How could I decide? Which would I enjoy most? Which would be *best*? One day I suddenly realised that 'I' didn't have to decide. I sat there, paying attention. The lights changed, a foot pressed the pedal, a hand changed gear, and the choice was made. I certainly never went straight on into the stone wall or bang into another car. And whichever way I went was fine. As time went on I found that more and more decisions were like this. It brought a great sense of freedom to let so many decisions alone.

You do not *have* to try to do anything or agonise about any decision. Let us suppose you are in the bath and the water is beginning to get cold. Do you get out now, or snuggle under the water a bit longer? Er . . . um. This is a trivial decision but, like getting out of bed in the morning, can colour your life. Knowing there is no real self to choose and no free will, you can only reflect that this body either will or will not get up, and indeed it does. Getting up decisively turns out not to be a matter of self-control and will-power, but of letting the false self get out of the way, and the decisions make themselves. The same is true of more complicated decisions; the brain may turn over the possibilities, argue the case one way or the other, come down on one side or other, but all of this can be done without, in addition, the false idea that someone inside is doing it. Rather the whole process seems to do itself.

Desires and hopes and preferences are probably the most difficult to deal with – I hope he'll get here in time, I must pass that exam, I hope I'll live to a ripe old age and get rich and famous, I want the *strawberry* one. All these hopes and desires are based on the idea of an inner self who must be kept happy, and their occurrence feeds the selfplex. So one trick is just to meet them all with a refusal to get involved. If there is no self then there is no point hoping or wishing for things for the sake of someone who does not exist. All these things are in another moment, not now. They do not matter when there is no one for them to matter to. Life really is possible without hope.

The result of this way of living seems somewhat counter-intuitive; that people become more decisive rather than less. On a second look this is not so surprising after all. From the memetic point of view the selfplex is not there to make the decisions, or for the sake of your happiness, or to make your life easier; it is there for the propagation of the memes that make it up. Its demolition allows more spontaneous and appropriate action. Clever thinking brains, installed with plenty of memes, are quite capable of making sound decisions without a selfplex messing them up.

A terrifying thought now raises its head. If I live by this kind of truth – without a self that takes responsibility for its actions, then what of morality? Surely, some would say, this kind of living is a recipe for selfishness and wickedness, for immorality and disaster. Well is it? One of the effects of this way of living is that you stop inflicting your own desires on the world around you and on the people you meet. This alone can mean quite a transformation.

Claxton describes the effect of giving up the illusion of a self in control. 'The thing that doesn't happen, but of which people are quite reasonably scared, is that I get worse. A common elaboration of the belief that control is real . . . is that I can, and must control "myself", and that unless I do, base urges will spill out and I will run amok.' Luckily, he goes on, the premise is false. 'So the dreaded mayhem does not happen. I do not take up wholesale rape and pillage and knocking down old ladies just for fun.' (Claxton 1986, p. 69). Instead, guilt, shame, embarrassment, self-doubt, and fear of failure ebb away and I become, contrary to expectation, a better neighbour.

In fact, we could reasonably have had faith in this from our understanding of memetics and of meme-driven altruism. Also, if it is true that the inner self is a memeplex and its control is illusory, then surely living a lie cannot be morally superior to accepting the truth. But if the self is a memeplex and can be dismantled, then what is left when it is gone? There is a human being, body, brain and memes, that behaves according to the environment it finds itself in and the memes it comes across. We know that the genes are responsible for much moral behaviour – they brought about kin-selection and reciprocal altruism, love of one's children, one's partners, and one's friends. And the memes are responsible for other kinds of sharing and caring. These behaviours will all still go on whether or not there is a selfplex cluttering the mind up as well.

Indeed, the selfplex can be blamed for much of the trouble. By its very nature the selfplex brings about self-recrimination, self-doubt, greed, anger, and all sorts of destructive emotions. When there is no selfplex, there is no concern about the future of my inner self – whether people like me or whether I did the 'right' thing or not – because there is no real 'I' to care about. This lack of self-concern means that you (the physical person) are free to notice other people more. Compassion and empathy come naturally. It is easy to see what another person needs, or how to act in a given situation, if there is no concern about a mythical self to get in the way. Perhaps the greater part of true morality is simply stopping all the harm that we normally do, rather than taking on any great and noble deeds; that is, the harm that comes from having a false sense of self.

Memetics thus brings us to a new vision of how we might live our lives. We can carry on our lives as most people do, under the illusion that there is a persistent conscious self inside who is in charge, who is responsible for my actions and who makes me me. Or we can live as human beings, body, brain, and memes, living out our lives as a complex interplay of replicators and environment, in the knowledge that that is all there is. Then we are no longer victims of the selfish selfplex. In this sense we can be truly free – not because we can rebel against the tyranny of the selfish replicators but because we know that there is no one to rebel.

References

Alexander, R. (1979). *Darwinism and Human Affairs*, Seattle, WA, University of Washington Press.

Allison, P. D. (1992). The cultural evolution of beneficent norms. *Social Forces*, 71, 279–301.

Ashby, R. (1960). *Design for a Brain*. New York, Wiley.

Baars, B. J. (1997). *In the Theatre of Consciousness: The Workspace of the Mind*. New York, Oxford University Press.

Bailey, L. W. and Yates, J. (eds.) (1996). *The Near-death Experience: A Reader*. New York/London, Routledge.

Baker, M. C. (1996). Depauperate meme pool of vocal signals in an island population of singing honeyeaters. *Animal Behaviour*, **51**, 853–8.

Baker, R. R. (1996). *Sperm Wars: Infidelity, Sexual Conflict and other Bedroom Battles*. London, Fourth Estate.

Baker, R. R. and Bellis, M. A. (1994). *Human Sperm Competition: Copulation, Masturbation, and Infidelity*. London, Chapman and Hall.

Baldwin, J. M. (1896). A new factor in evolution. *American Naturalist*, **30**, 441–51, 536–53.

Baldwin, J. M. (1909). *Darwin and the Humanities*, Baltimore, MD, Review Publishing.

Ball, J. A. (1984), Memes as replicators. *Ethology and Sociobiology*, **5**, 145–61.

Bandura, A. and Walters, R. H. (1963). *Social Learning and Personality Development*. New York, Holt, Rinehart & Winston.

Barkow, J. H., Cosmides, L. and Tooby, J. (eds.) (1992). *The Adapted Mind: Evolutionary Psychology and the Generation of Culture*. New York, Oxford University Press.

Barrett, S. and Jarvis, W. T. (eds.) (1993). *The Health Robbers: A Close Look at Quackery in America*. Buffalo, NY, Prometheus.

Bartlett, F. C. (1932). *Remembering: A Study in Experimental and Social Psychology*. Cambridge University Press.

Barton, R. A. and Dunbar, R. I. M. (1997). Evolution of the social brain. In *Machiavellian Intelligence: II. Extensions and Evaluations*, (ed. A. Whiten and R. W. Byrne), pp. 240–63. Cambridge University Press.

Basalla, G. (1988). *The Evolution of Technology*. Cambridge University Press.

Batchelor, S. (1994). *The Awakening of the West: The Encounter of Buddhism and Western Culture*. London, HarperCollins.

Batson, C. D. (1995). Prosocial motivation: Why do we help others? In

Advanced Social Psychology, (ed. A. Tesser), pp. 333–81. New York, McGraw-Hill.

Bauer, G. B. and Johnson, C. M. (1994). Trained motor imitation by bottlenose dolphins (*Tursiops truncatus*). *Perceptual and Motor Skills*, **79**, 1307–15.

Benor, D. J. (1994). *Healing Research: Holistic Energy, Medicine and Spirituality*. Munich, Helix.

Benzon, W. (1996). Culture as an evolutionary arena. *Journal of Social and Evolutionary Systems*, **19**, 321–62.

Berlin, B. and Kay, P. (1969). *Basic Color Terms: Their Universality and Evolution*. Berkeley, CA, University of California Press.

Bickerton, D. (1990). *Language and Species*. Chicago, IL, University of Chicago Press.

Bikhchandani, S., Hirshleifer, D. and Welch, I. (1992). A theory of fads, fashion, custom and cultural change as informational cascades. *Journal of Political Economy*, **100**, 992–1026.

Blackmore, S. J. (1993). *Dying to Live: Science and the Near Death Experience*. Buffalo, NY, Prometheus.

Blackmore, S. J. (1995). Paying attention. *New Ch'an Forum*, No. 12, 9–15.

Blackmore, S. J. (1997). Probability misjudgment and belief in the paranormal: a newspaper survey. *British Journal of Psychology*, **88**, 683–9.

Blackmore, S. J. (in press). Waking from the Meme Dream. In *The Psychology of Awakening: Buddhism, Science and Psychotherapy*, (ed. G. Watson, G. Claxton and S. Batchelor). Dorset, Prism.

Blackmore, S. J. and Troscianko, T. (1985). Belief in the paranormal: Probability judgements, illusory control, and the chance baseline shift. *British Journal of Psychology*, **76**, 459–68.

Blackmore, S. J., Brelstaff, G., Nelson, K. and Troscianko, T. (1995). Is the richness of our visual world an illusion? Transsaccadic memory for complex scenes. *Perception*, **24**, 1075–81.

Blakemore, C. and Greenfield, S. (eds.) (1987). *Mindwaves*. Oxford, Blackwell.

Bonner, J. T. (1980). *The Evolution of Culture in Animals*. Princeton, NJ, Princeton University Press.

Bowker, J. (1995). *Is God a Virus?* London, SPCK.

Boyd, R. and Richerson, P. J. (1985). *Culture and the Evolutionary Process*. Chicago, IL, University of Chicago Press.

Boyd, R. and Richerson, P. J. (1990). Group selection among alternative evolutionarily stable strategies. *Journal of Theoretical Biology*, **145**, 331–42.

Brodie, R. (1996). *Virus of the Mind: The New Science of the Meme*. Seattle, WA, Integral Press.

Bucke, R. M. (1901). *Cosmic Consciousness: A Study in the Evolution of the Human Mind*. (London, Arkana, Penguin, 1991.)

Buss, D. M. (1994). *The Evolution of Desire: Strategies of Human Mating*. New York, Basic Books.

Byrne, R. W. and Whiten, A. (eds.) (1988). *Machiavellian Intelligence: Social*

Expertise and the Evolution of Intellect in Monkeys, Apes and Humans. Oxford University Press.

Call, J. and Tomasello, M. (1995). Use of social information in the problem solving of orangutans (*Pongo pygmaeus*) and human children (*Homo sapiens*). *Journal of Comparative Psychology*, **109**, 308–20.

Calvin, W. (1987). The brain as a Darwin machine. *Nature*, **330**, 33–44.

Calvin, W. (1996). *How Brains Think*, London, Phoenix.

Campbell, D. T. (1960). Blind variation and selective retention in creative thought as in other knowledge processes. *Psychological Review*, **67**, 380–400.

Campbell, D. T. (1965). Variation and selective retention in sociocultural evolution. In *Social Change in Developing Areas: A reinterpretation of evolutionary theory* (ed. H. R. Barringer, G. L. Blanksten and R. W. Mack), pp. 19–49. Cambridge, MA, Schenkman.

Campbell, D. T. (1974). Evolutionary epistemology. In *The Philosophy of Karl Popper*, Vol. 1, (ed. P. A. Schlipp), pp. 413–63. La Salle, IL, Open Court Publishing.

Campbell, D. T. (1975). On the conflicts between biological and social evolution and between psychology and moral tradition. *American Psychologist*, **30**, 1103–26.

Carlson, N. R. (1993). *Psychology: The Science of Behavior*, (4th edn). Boston, MA, Allyn & Bacon.

Cavalli-Sforza, L. L. and Feldman, M. W. (1981). *Cultural Transmission and Evolution: A Quantitative Approach*. Princeton, NJ, Princeton University Press.

Chagnon, N. A. (1992). *Yanomamö*, (4th edn). New York, Harcourt Brace Jovanovich.

Chalmers, D. (1996). *The Conscious Mind*. Oxford University Press.

Cheney, D. L. and Seyfarth, R. M. (1990). The representation of social relations by monkeys. *Cognition*, **37**, 167–96.

Churchland, P. S. (1998). Brainshy: Nonneural theories of conscious experience. In *Toward a Science of Consciousness: The Second Tucson Discussions and Debates*, (ed. S. R. Hameroff, A. W. Kaszniak and A. C. Scott), pp. 109–26. Cambridge, MA, MIT Press.

Churchland, P. S. and Sejnowski, T. J. (1992). *The Computational Brain*. Cambridge, MA, MIT Press.

Cialdini, R. B. (1994). *Influence: The Psychology of Persuasion*. New York, Morrow.

Cialdini, R. B. (1995). The principles and techniques of social influence. In *Advanced Social Psychology*, (ed. A. Tesser), pp. 257–81. New York, McGraw-Hill.

Claxton, G. (ed.) (1986). *Beyond Therapy: The Impact of Eastern Religions on Psychological Theory and Practice*. London, Wisdom. (Dorset, Prism, 1996.)

Claxton, G. (1994). *Noises from the Darkroom*. London, Aquarian.

Cloak, F. T. (1975). Is a cultural ethology possible? *Human Ecology*, **3**, 161–82.

Conlisk, J. (1980). Costly optimizers versus cheap imitators. *Journal of Economic Behavior and Organization*, 1, 275–93.

Crick, F. (1994). *The Astonishing Hypothesis: The Scientific Search for the Soul.* New York, Charles Scribner's Sons.

Cronin, H. (1991). *The Ant and the Peacock.* Cambridge University Press.

Crook, J. H. (1980). *The Evolution of Human Consciousness.* Oxford University Press.

Crook, J. H. (1989). Socioecological paradigms, evolution and history: perspectives for the 1990s. In *Comparative Socioecology*, (ed. V. Standen and R. A. Foley). Oxford, Blackwell.

Crook, J. H. (1995). Psychological processes in cultural and genetic coevolution. In *Survival and Religion: Biological Evolution and Cultural Change*, (ed. E. Jones and V. Reynolds), pp. 45–110. London, Wiley.

Csikszentmihalyi, M. (1990). *Flow: The Psychology of Optimal Experience.* New York, Harper & Row.

Csikszentmihalyi, M. (1993). *The Evolving Self: A Psychology for the Third Millennium.* New York, HarperCollins.

Damasio, A. (1994). *Descartes' Error: Emotion, Reason and the Human Brain.* New York, Putnam.

Darwin, C. (1859). *On the Origin of Species by Means of Natural Selection.* London, Murray. (London, Penguin, 1968).

Darwin, C. (1871). *The Descent of Man and Selection in Relation to Sex.* London, John Murray.

Dawkins, R. (1976). *The Selfish Gene.* Oxford University Press. (Revised edition with additional material, 1989.)

Dawkins, R. (1982). *The Extended Phenotype.* Oxford, Freeman.

Dawkins, R. (1986). *The Blind Watchmaker.* Harlow, Essex, Longman.

Dawkins, R. (1993). Viruses of the mind. In *Dennett and his Critics: Demystifying Mind*, (ed. B. Dahlbohm), pp. 13–27. Oxford, Blackwell.

Dawkins, R. (1994). Burying the vehicle. *Behavioral and Brain Sciences*, 17, 616–17.

Dawkins, R. (1996*a*). *Climbing Mount Improbable.* London, Penguin.

Dawkins, R. (1996*b*). Mind viruses. In *Ars Electronica Festival 1996: Memesis: The Future of Evolution* (ed. G. Stocker and C. Schöpf), pp. 40–7, Vienna, Springer.

Deacon, T. (1997). *The Symbolic Species: The Co-evolution of Language and the Human Brain.* London, Penguin.

Dean, G., Mather, A. and Kelly, I. W. (1996). Astrology. In *The Encyclopedia of the Paranormal*, (ed. G. Stein), pp. 47–99. Buffalo, New York, Prometheus.

Delius, J. (1989). Of mind memes and brain bugs, a natural history of culture. In *The Nature of Culture.* (ed. W. A. Koch), pp. 26–79. Bochum, Germany, Bochum Publications.

Dennett, D. (1978). *Brainstorms: Philosophical Essays on Mind and Psychology.* Montgomery, VT, Bradford Books.

Dennett, D. (1984). *Elbow Room: The Varieties of Free Will Worth Wanting.* Cambridge, MA., Bradford Books.

Dennett, D. (1991). *Consciousness Explained.* Boston, MA, Little Brown.

Dennett, D. (1995). *Darwin's Dangerous Idea.* London, Penguin.

Dennett, D. (1997). *The evolution of evaluators.* Paper presented at the International School of Economic Research, Siena.

Dennett, D. (1998). Personal communication (Dennett suggested the terms 'meme-fountain' and 'meme-sink').

Descartes, R. (1641). *Discourse on Method and the Meditations.* (London, Penguin, 1968.)

Diamond, J. (1997). *Guns, Germs and Steel.* London, Cape.

Donald, M. (1991). *Origins of the Modern Mind: Three Stages in the Evolution of Culture and Cognition.* Cambridge, MA, Harvard University Press.

Donald, M. (1993). *Précis of* Origins of the modern mind: Three stages in the evolution of culture and cognition. *Behavioral and Brain Sciences*, **16**, 737–91. (with commentaries by others.)

Dossey, L. (1993). *Healing Words: The Power of Prayer and the Practice of Medicine.* San Francisco, CA, HarperCollins.

Dunbar, R. (1996). *Grooming, Gossip and the Evolution of Language.* London, Faber and Faber.

Durham, W. H. (1991). *Coevolution: Genes, Culture and Human Diversity.* Stanford, CA, Stanford University Press.

Du Preez, P. (1996). The evolution of altruism: A brief comment on Stern's 'Why do people sacrifice for their nations?' *Political Psychology*, **17**, 563–7.

Edelman, G. M. (1989). *Neural Darwinism: The Theory of Neuronal Group Selection.* Oxford University Press.

Eisenberg, D. M., Kessler, R. C., Foster, C., Norlock, F. E., Calkins, D. R. and Delbanco, T. L. (1993). Unconventional medicine in the United States. *New England Journal of Medicine*, **328**, 246–52.

Eagly, A. H. and Chaiken, S. (1984). Cognitive theories of persuasion, In *Advances in Experimental Social Psychology*, Vol. 17, (ed. L. Berkowitz), pp. 267–359. New York, Academic Press.

Ernst, E. (1998). The rise and fall of complementary medicine. *Journal of the Royal Society of Medicine*, **91**, 235–6.

Festinger, L. (1957). *A Theory of Cognitive Dissonance.* Stanford, CA, Stanford University Press.

Fisher, J. and Hinde, R. A. (1949). The opening of milk bottles by birds. *British Birds*, **42**, 347–57.

Fisher, R. A. (1930). *The Genetical Theory of Natural Selection.* Oxford University Press.

Forer, B. R. (1949). The fallacy of personal validation: A classroom demonstration of gullibility. *Journal of Abnormal and Social Psychology*, **44**, 118–23.

Freeman, D. (1996). *Margaret Mead and the Heretic: The Making and Unmaking of an Anthropological Myth.* London, Penguin.

Gabora, L. (1997). The origin and evolution of culture and creativity. *Journal of Memetics*, **1**, http://www.cpm.mmu.ac.uk/jom-emit/1997/vol 1/gabora_l.html.

Galef, B. G. (1992). The question of animal culture. *Human Nature*, **3**, 157–78.

Gallup, G. H. and Newport F. (1991). Belief in paranormal phenomena among adult Americans. *Skeptical Inquirer*, **15**, 137–46.

Gatherer, D. (1997). The evolution of music – a comparison of Darwinian and dialectical methods. *Journal of Social and Evolutionary Systems*, **20**, 75–93.

Gatherer, D. (1998). Meme pools, World 3, and Averroës's vision of immortality. *Zygon*, **33**, 203–19.

Gould, S. J. (1979). Shades of Lamarck. *Natural History*, **88**, 22–8.

Gould, S. J. (1991). *Bully for Brontosaurus.* New York, Norton.

Gould, S. J. (1996*a*). *Full House.* New York, Harmony Books. (Published in the UK as *Life's Grandeur*, London, Cape.)

Gould, S. J. (1996*b*). BBC Radio 4. *Start the Week* Debate with S. Blackmore, S. Fry and O. Sacks, 11 November.

Gould, S. J. and Lewontin, R. (1979). The spandrels of San Marco and the Panglossian paradigm: A critique of the adaptationist programme. *Proceedings of the Royal Society*, **B205**, 581–98.

Grant, G. (1990). Memetic lexicon. http://pespmc1.vub.ac.be/*memes.html.

Gregory, R. L. (1981). *Mind in Science: A History of Explanations in Psychology and Physics.* London, Weidenfeld & Nicolson.

Grosser, D., Polansky, N. and Lippitt, R. (1951). A laboratory study of behavioral contagion. *Human Relations*, **4**, 115–42.

Hameroff, S. R. (1994). Quantum coherence in microtubules: A neural basis for emergent consciousness? *Journal of Consciousness Studies*, **1**, 91–118.

Hamilton, W. D. (1963). The evolution of altruistic behaviour. *American Naturalist*, **97**, 354–6.

Hamilton, W. D. (1964). The genetical evolution of social behaviour: 1. *Journal of Theoretical Biology*, **7**, 1–16.

Hamilton, W. D. (1996). *Narrow Roads of Gene Land: 1. The Evolution of Social Behaviour.* Oxford, Freeman/Spektrum.

Hartung, J. (1995). Love thy neighbour: the evolution of in-group morality. *Skeptic*, **3:4**, 86–99.

Harvey, P. H. and Krebs, J. R. (1990). Comparing brains. *Science*, **249**, 140–6.

Heyes, C. M. (1993). Imitation, culture and cognition. *Animal Behaviour*, **46**, 999–1010.

Heyes, C. M. and Galef, B. G. (ed.) (1996). *Social Learning in Animals: The Roots of Culture.* San Diego, CA, Academic Press.

Hofstadter, D. R. (1985). *Metamagical Themas: Questing for the Essence of Mind and Pattern.* New York, Basic Books.

Hull, D. L. (1982). The naked meme. In *Learning, Development and Culture*, (ed. H. C. Plotkin), pp. 273–327. London, Wiley.

Hull, D. L. (1988*a*). Interactors versus vehicles. In *The Role of Behaviour in Evolution*, (ed. H. C. Plotkin), pp. 19–50. Cambridge, MA, MIT Press.

Hull, D. L. (1988*b*). A mechanism and its metaphysic: an evolutionary account of the social and conceptual development of science. *Biology and Philosophy*, **3**, 123–55.

Hume, D. (1739–40). *A Treatise of Human Nature*. Oxford.

Humphrey, N. (1986). *The Inner Eye*. London, Faber and Faber.

Humphrey, N. (1995). *Soul Searching: Human Nature and Supernatural Belief*. London, Chatto & Windus.

Jacobs, D. M. (1993). *Secret Life: First hand accounts of UFO abductions*. London, Fourth Estate.

Jerison, H. J. (1973). *Evolution of the Brain and Intelligence*. New York, Academic Press.

Johnson, T. R. (1995). The significance of religion for aging well. *American Behavioral Scientist*, **39**, 186–209.

Kauffman, S. (1995). *At Home in the Universe: The Search for Laws of Complexity*. Oxford University Press.

King, M., Speck, P. and Thomas, A. (1994). Spiritual and religious beliefs in acute illness – is this a feasible area for study? *Social Science and Medicine*, **38**, 631–6.

Krings, M., Stone, A., Schmitz, R. W., Krainitzki, H., Stoneking, M. and Pääbo, S. (1997).Neanderthal DNA sequences and the origin of modern humans. *Cell*, **90**, 19–30.

Langer, E. J. (1975). The illusion of control. *Journal of Personality and Social Psychology*, **32**, 311–28.

Leakey, R. (1994). *The Origin of Humankind*. London, Weidenfeld & Nicolson.

Levy, D. A. and Nail, P. R. (1993). Contagion: A theoretical and empirical review and reconceptualization. *Genetic, Social, and General Psychology Monographs*. **119**, 235–84.

Libet, B. (1981). The experimental evidence of subjective referral of a sensory experience backwards in time. *Philosophy of Science*, **48**, 182–97.

Libet, B. (1985). Unconscious cerebral initiative and the role of conscious will in voluntary action. *Behavioral and Brain Sciences*, **8**, 529–39. (With commentaries 539–66; and *BBS*, **10**, 318–21.)

Libet, B., Pearl, D. K., Morledge, D. E., Gleason, C. A. Hosobuchi, Y. and Barbaro, N. M. (1991). Control of the transition from sensory detection to sensory awareness in man by the duration of a thalamic stimulus: The cerebral 'time-on' factor. *Brain*, **114**, 1731–57.

Lumsden, C. J. and Wilson, E. O. (1981). *Genes, Mind and Culture*. Cambridge, MA, Harvard University Press.

Lynch, A. (1991). Thought contagion as abstract evolution. *Journal of Ideas*, **2**, 3–10.

Lynch, A. (1996). *Thought Contagion: How Belief Spreads through Society*. New York, Basic Books.

Lynch, A., Plunkett, G. M., Baker, A. J. and Jenkins, P. F. (1989). A model of cultural evolution of chaffinch song derived with the meme concept. *The American Naturalist*, **133**, 634–53.

Machiavelli, N. (*c*.1514). *The Prince*. (London, Penguin, 1961, trans. G. Bull.)

Mack, J. E. (1994). *Abduction: Human encounters with aliens*. London, Simon & Schuster.

Mackay, C. (1841). *Extraordinary Popular Delusions and the Madness of Crowds*. (Reprinted, New York, Wiley, 1996.)

Marsden, P. (1997). *Crash contagion and the Death of Diana: Memetics as a new paradigm for understanding mass behaviour*. Paper presented at the conference 'Death of Diana', University of Sussex, 14 November.

Marsden, P. (1998*a*). Memetics as a new paradigm for understanding and influencing customer behaviour. *Marketing Intelligence and Planning*, **16**, 363–8.

Marsden, P. (1998*b*). *Operationalising memetics: suicide, the Werther Effect, and the work of David P. Phillips*. Paper presented at the Fifteenth International Congress on Cybernetics, Symposium on Memetics, Namur, August.

Marx, K. (1904). *A Contribution to the Critique of Political Economy*. Chicago, IL, Charles H. Kerr.

Maynard Smith, J. (1996). Evolution – natural and artificial. In *The Philosophy of Artificial Life*, (ed. M. A. Boden), pp. 173–8. Oxford University Press.

Maynard Smith, J. and Szathmáry, E. (1995). *The Major Transitions of Evolution*. Oxford, Freeman/Spektrum.

Mead, M. (1928). *Coming of Age in Samoa*. (London, Penguin, 1963.)

Meltzoff, A. N. (1988). Imitation, objects, tools, and the rudiments of language in human ontogeny. *Human Evolution*, **3**, 45–64.

Meltzoff, A. N. (1990). Towards a developmental cognitive science: the implications of cross-modal matching and imitation for the development of representation and memory in infancy. *Annals of the New York Academy of Science*, **608**, 1–37.

Meltzoff, A. N. (1996). The human infant as imitative generalist: A 20-year progress report on infant imitation with implications for comparative psychology. In *Social Learning in Animals: The Roots of Culture*, (ed. C. M. Heyes and B. G. Galef), pp. 347–70, San Diego, CA, Academic Press.

Meltzoff, A. N. and Moore, M. K. (1977). Imitation of facial and manual gestures by human neonates. *Science*, **198**, 75–8.

Mestel, R. (1995). Arts of seduction. *New Scientist*, 23/30 December, 28–31.

Midgley, M. (1994). Letter to the Editor. *New Scientist*, 12 February, 50.

Miller, G. (1993). Evolution of the Human Brain through Runaway Sexual Selection. PhD thesis, Stanford University Psychology Department.

Miller, G. (1998). How mate choice shaped human nature: A review of sexual selection and human evolution. In *Handbook of Evolutionary Psychology: Ideas, Issues, and Applications* (ed. C. Crawford and D. Krebs), pp. 87–129, Mahwah, NJ: Erlbaum.

Miller, N. E. and Dollard, J. (1941). *Social Learning and Imitation.* New Haven, CT, Yale University Press.

Mithen, S. (1996). *The Prehistory of the Mind.* London, Thames and Hudson.

Moghaddam, F. M., Taylor, D. M. and Wright, S. C. (1993). *Social Psychology in Cross-Cultural Perspective.* New York, Freeman.

Myers, F. W. H. (1903). *Human Personality and its Survival of Bodily Death.* London, Longmans, Green.

Osis, K. and Haraldsson, E. (1977). Deathbed observations by physicians and nurses: A cross-cultural survey *Journal of the American Society for Psychical Research*, 71, 237–59.

Otero, C. P. (1990). The emergence of *homo loquens* and the laws of physics. *Behavioral and Brain Sciences*, **13**, 747–50.

Parfit, D. (1987). Divided minds and the nature of persons. In *Mindwaves*, (ed. C. Blakemore and S. Greenfield), pp. 19–26. Oxford, Blackwell.

Penrose, R. (1994). *Shadows of the Mind: A Search for the Missing Science of Consciousness.* Oxford University Press.

Persinger, M. A. (1983) Religious and mystical experiences as artifacts of temporal lobe function: A general hypothesis. *Perceptual and Motor Skills*, 57, 1255–62.

Phillips, D. P. (1980). Airplane accidents, murder, and the mass media: Towards a theory of imitation and suggestion. *Social Forces*, **58**, 1000–24.

Pinker, S. (1994). *The Language Instinct.* New York, Morrow.

Pinker, S. (1998). *How the Mind Works.* London, Penguin.

Pinker, S. and Bloom, P. (1990). Natural language and natural selection. *Behavioral and Brain Sciences*, **13**, 707–84. (with commentaries by others.)

Plimer, I. (1994). *Telling Lies for God.* Milsons Point, NSW, Australia, Random House.

Plotkin, H. C. (ed.) (1982). *Learning, Development and Culture: Essays in Evolutionary Epistemology.* Chichester, Wiley.

Plotkin, H. C. (1993). *Darwin Machines and the Nature of Knowledge.* London, Penguin.

Popper, K. R. (1972). *Objective Knowledge: An Evolutionary Approach.* Oxford University Press.

Popper, K. R. and Eccles, J. C. (1977). *The Self and its Brain: An Argument for Interactionism.* Berlin, Springer.

Provine, R. R. (1996). Contagious yawning and laughter: Significance for sensory feature detection, motor pattern generation, imitation, and the evolution of social behaviour. In *Social Learning in Animals: The Roots of Culture*, (ed. C. M. Heyes and B. G. Galef), pp. 179–208. San Diego, CA, Academic Press.

Pyper, H. S. (1998). The selfish text: the Bible and memetics. In *Biblical Studies and Cultural Studies*, (ed. J. C. Exum and S. D. Moore), pp. 70–90. Sheffield Academic Press.

Reiss, D. and McCowan, B. (1993). Spontaneous vocal mimicry and production by bottlenose dolphins (*Tursiops truncatus*): Evidence for vocal learning. *Journal of Comparative Psychology*, **107**, 301–12.

Richerson, P. J. and Boyd, R. (1989). The role of evolved predispositions in cultural evolution: Or, human sociobiology meets Pascal's wager. *Ethology and Sociobiology*, **10**, 195–219.

Richerson, P. J. and Boyd, R. (1992). Cultural inheritance and evolutionary ecology. In *Evolutionary Ecology and Human Behaviour*, (ed. E. A. Smith and B. Winterhalder), pp. 61–92. Chicago, IL, Aldine de Gruyter.

Ridley, Mark (1996). *Evolution*, (2nd edn). Oxford, Blackwell.

Ridley, Matt (1993). *The Red Queen: Sex and the Evolution of Human Nature*. London, Viking.

Ridley, Matt (1996). *The Origins of Virtue*. London, Viking.

Ring, K. (1992). *The Omega Project*. New York, Morrow.

Rose, N. J. (1997). Personal communication.

Rose, N. J. (1998). Controversies in meme theory. *Journal of Memetics: Evolutionary Models of Information Transmission*, **2**, http://www.cpm.mmu.ac.uk/jom-emit/1998/vol 2/rose_n.html.

Runciman, W. G. (1998). The selectionist paradigm and its implications for sociology. *Sociology*, **32**, 163–88.

Sheils, D. (1978). A cross-cultural study of beliefs in out-of-the-body experiences. *Journal of the Society for Psychical Research*, **49**, 697–741.

Sherry, D. F. and Galef, B. G. (1984). Cultural transmission without imitation: milk bottle opening by birds. *Animal Behaviour*, **32**, 937–8.

Showalter, E. (1997). *Hystories: Hysterical Epidemics and Modern Culture*. New York, Columbia University Press.

Silver, L. M. (1998). *Remaking Eden: Cloning and Beyond in a Brave New World*. London, Weidenfeld & Nicolson.

Singh, D. (1993). Adaptive significance of female physical attractiveness: role of waist-to-hip ratio. *Journal of Personality and Social Psychology*, **65**, 293–307.

Skinner, B. F. (1953). *Science and Human Behavior*. New York, Macmillan.

Spanos, N. P., Cross, P. A., Dickson, K., and DuBreuil, S. C. (1993). Close encounters: An examination of UFO experiences. *Journal of Abnormal Psychology*, **102**, 624–32.

Speel, H.-C. (1995). *Memetics: On a conceptual framework for cultural evolution*. Paper presented at the symposium 'Einstein meets Magritte', Free University of Brussels, June.

Sperber, D. (1990). The epidemiology of beliefs. In *The Social Psychological Study of Widespread Beliefs*, (ed. C. Fraser and G. Gaskell), pp. 25–44. Oxford Univesity Press.

Stein, G. (ed.) (1996). *The Encyclopedia of the Paranormal*. Buffalo, NY, Prometheus.

Strawson, G. (1997). The self. *Journal of Consciousness Studies*, **4**, 405–28.

Symons, D. (1979). *The Evolution of Human Sexuality*. New York, Oxford University Press.

Thorndike, E. L. (1898). Animal intelligence: An experimental study of the associative processes in animals. *Psychological Review Monographs*, 2, No. 8.

Tomasello, M., Kruger, A. C. and Ratner, H. H. (1993). Cultural learning. *Behavioral and Brain Sciences*, **16**, 495–552.

Tooby, J. and Cosmides, L. (1992). The psychological foundations of culture. In *The Adapted Mind: Evolutionary Psychology and the Generation of Culture*, (ed. J. H. Barkow, L. Cosmides and J. Tooby), pp. 19–136. New York, Oxford University Press.

Toth, N. and Schick, K. (1993). Early stone industries and inferences regarding language and cognition. In *Tools, Language and Cognition in Human Evolution* (ed. K. Gibson and T. Ingold), pp. 346–62. Cambridge University Press.

Trivers, R. L. (1971). The evolution of reciprocal altruism. *Quarterly Review of Biology*, **46**, 35–56.

Trivers, R. L. (1972). Parental investment and sexual selection. In *Sexual Selection and the Descent of Man*, (ed. B. Campbell), pp. 136–79. Chicago, IL, Aldine de Gruyter.

Trivers, R. L. (1985). *Social Evolution*. Menlo Park, CA, Benjamin/Cummings.

Tudge, C. (1995). *The Day before Yesterday: Five Million Years of Human History*. London, Cape.

Turkle, S. (1995). *Life on the Screen: Identity in the Age of the Internet*. New York, Simon & Schuster.

Ulett, G. (1992). *Beyond Yin and Yang: How Acupuncture Really Works*. St. Louis, MO, Warren H. Green.

Ulett, G. A., Han, S. and Han, J. (1998). Electroacupuncture: Mechanisms and clinical application. *Biological Psychiatry*, **44**, 129–38.

Wagstaff, G. F. (1998). Equity, justice and altruism. *Current Psychology*, **17**, 111–34.

Walker, A. and Shipman, P. (1996). *The Wisdom of Bones: In Search of Human Origins*. London, Weidenfeld & Nicolson.

Wallace, A. R. (1891). *Natural Selection and Tropical Nature: Essays on Descriptive and Theoretical Biology*. London, Macmillan.

Warraq, I. (1995). *Why I am not a Muslim*. Buffalo, NY, Prometheus.

Watson, J. D. (1968). *The Double Helix*. London, Weidenfeld & Nicolson.

Whiten, A. and Byrne, R. W. (1997). *Machiavellian Intelligence: II. Extensions and Evaluations*. Cambridge University Press.

Whiten, A. Custance, D. M., Gomez, J.-C., Teixidor, P. and Bard, K. A. (1996). Imitative learning of artificial fruit processing in children (*Homo sapiens*) and chimpanzees (*Pan troglodytes*). *Journal of Comparative Psychology*, **110**, 3–14.

Whiten, A. and Ham, R. (1992). On the nature and evolution of imitation in the animal kingdom: Reappraisal of a century of research. In *Advances in the Study of Behavior*, Vol. 21, (ed. P. J. B. Slater, J. S. Rosenblatt, C. Beer and M. Milinski), pp. 239–81. San Diego, CA, Academic Press.

Williams, G. C. (1966). *Adaptation and Natural Selection*. Princeton, NJ, Princeton University Press.

Wills, C. (1993). *The Runaway Brain: The Evolution of Human Uniqueness.* New York, Basic Books.

Wilson, D. S. and Sober, E. (1994). Reintroducing group selection to the human behavioral sciences. *Behavioral and Brain Sciences*, **17**, 585–654 (with commentaries by others).

Wilson, E. O. (1978). *On Human Nature.* Cambridge, MA, Harvard University Press.

Wilson, I. (1987). *The After Death Experience.* London, Sidgwick & Jackson.

Wispé, L. G. and Thompson, J. N. (1976). The war between the words: biological versus social evolution and some related issues. *American Psychologist*, **31**, 341–84.

Wright, D. (1998). *Translated terms as meme-products: The struggle for existence in Late Qing chemical terminologies.* Paper presented at the conference 'China and the West', Technical University of Berlin, August.

Wright, R. (1994). *The Moral Animal.* New York, Pantheon.

Yando, R., Seitz, V., and Zigler, E. (1978). *Imitation: A Developmental Perspective.* New York, Wiley.

Young, J. Z. (1965). *A Model of the Brain.* Oxford, Clarendon.

Zentall, T. R. and Galef, B. G. (ed.) (1988). *Social Learning: Psychological and Biological Perspectives.* Hillsdale, NJ, Erlbaum.

Index